DATE DUE

WAVES, TIDES AND SHALLOW-WATER PROCESSES

THE OCEANOGRAPHY COURSE TEAM

Authors
Joan Brown
Angela Colling
Dave Park
John Phillips
Dave Rothery
John Wright

Editor
Gerry Bearman

Design and Illustration
Sue Dobson
Ray Munns
Ros Porter
Jane Sheppard

This Volume forms part of an Open University course. For general availability of all the Volumes in the Oceanography Series, please contact your regular supplier, or in case of difficulty the appropriate Pergamon office.

Further information on Open University courses may be obtained from: The Admissions Office, The Open University, P.O. Box 48, Walton Hall, Milton Keynes, MK7 6AA.

Cover illustration: Satellite photograph showing distribution of phytoplankton pigments in the North Atlantic off the US coast in the region of the Gulf Stream and the Labrador Current. *(NASA and O. Brown and R. Evans, University of Miami.)*

WAVES, TIDES AND SHALLOW-WATER PROCESSES

PREPARED BY AN OPEN UNIVERSITY COURSE TEAM

PERGAMON PRESS
OXFORD · NEW YORK · BEIJING · FRANKFURT · SÃO PAULO · SYDNEY
TOKYO · TORONTO

in association with

THE OPEN UNIVERSITY
WALTON HALL, MILTON KEYNES, MK7 6AA, ENGLAND

First edition 1989

Library of Congress Cataloging in Publication Data
Waves, tides and shallow-water processes
1. Waves.   2. Tides.   I. Open University. Oceanography Course Team.
GC211.2.W38   1989   551.47—dc19   88–28868

British Library Cataloguing in Publication Data
Waves, tides and shallow-water processes
1. Oceans. Dynamics
I. Open University, *Oceanography Course Team*
551.47
ISBN 0-08-036372-5 Hardcover
ISBN 0-08-036371-7 Flexicover

Jointly published by the Open University, Walton Hall, Milton Keynes, MK7 6AA and Pergamon Press plc, Headington Hill Hall, Oxford OX3 0BW.

Designed by the Graphic Design Group of The Open University.

Printed in Great Britain by BPCC Wheatons Ltd., Exeter

CONTENTS

ABOUT THIS VOLUME   **4**

CHAPTER 1   **WAVES**

1.1	**WHAT ARE WAVES?**	**8**
1.1.1	Types of waves	8
1.1.2	Wind-generated waves on the ocean	10
1.1.3	The fully developed sea	12
1.1.4	Wave height and wave steepness	12
1.2	**WAVE-FORMS**	**14**
1.2.1	Motion of water particles	15
1.2.2	Surface wave theory	17
1.2.3	Wave speed in deep and in shallow water	17
1.2.4	Assumptions made in surface wave theory	19
1.3	**WAVE DISPERSION AND GROUP SPEED**	**19**
1.4	**WAVE ENERGY**	**21**
1.4.1	Propagation of wave energy	21
1.4.2	Swell	22
1.4.3	Attenuation of wave energy	24
1.4.4	Wave refraction	25
1.4.5	Waves approaching the shore	27
1.4.6	Waves breaking upon the shore	29
1.5	**WAVES OF UNUSUAL CHARACTER**	**31**
1.5.1	Waves and currents	31
1.5.2	Giant waves	32
1.5.3	Tsunamis	34
1.5.4	Seiches	35
1.6	**MEASUREMENT OF WAVES**	**37**
1.6.1	Satellite observations of waves	37
1.7	**SUMMARY OF CHAPTER 1**	**39**

CHAPTER 2   **TIDES**

2.1	**TIDE-PRODUCING FORCES—THE EARTH–MOON SYSTEM**	**43**
2.1.1	Variations in the lunar-induced tides	48
2.2	**TIDE-PRODUCING FORCES—THE EARTH–SUN SYSTEM**	**50**
2.2.1	Interaction of solar and lunar tides	50
2.3	**THE DYNAMIC THEORY OF TIDES**	**52**
2.3.1	Prediction of tides by the harmonic method	57
2.4	**TYPES OF TIDE**	**58**
2.4.1	Tides and tidal currents in shallow seas	59
2.4.2	Storm surges	61
2.4.3	Tides in rivers and estuaries	62
2.4.4	Tidal power	63
2.5	**SUMMARY OF CHAPTER 2**	**64**

CHAPTER 3 — AN INTRODUCTION TO SHALLOW-WATER ENVIRONMENTS AND THEIR SEDIMENTS

3.1	THE SUPPLY OF SEDIMENT TO SHALLOW-WATER ENVIRONMENTS	67
3.2	VARIATIONS IN THE SUPPLY AND DISTRIBUTION OF SHALLOW-WATER SEDIMENTS	69
3.3	THE RELATIONSHIPS BETWEEN SHALLOW-WATER ENVIRONMENTS AND THEIR CHANGES WITH TIME	70
3.4	SUMMARY OF CHAPTER 3	71

CHAPTER 4 — SEDIMENT MOVEMENT BY WAVES AND CURRENTS

4.1	FLUID FLOW	72
4.1.1	Frictional forces and the boundary layer	73
4.1.2	The flow of water in the boundary layer	74
4.1.3	The significance of the viscous sublayer	75
4.1.4	Current shear at the bed	76
4.1.5	Roughness length	78
4.1.6	Velocity profiles in the sea	80
4.2	SEDIMENT EROSION	81
4.2.1	Erosion of non-cohesive sediments	82
4.2.2	Erosion of cohesive sediments	83
4.3	THE RATE OF SEDIMENT TRANSPORT	84
4.3.1	The bedload transport rate	84
4.3.2	The rate of transport of the suspended load	86
4.4	THE DEPOSITION OF SEDIMENT	87
4.4.1	Deposition of the bedload	87
4.4.2	Deposition of the suspended load	87
4.4.3	Rates of sediment deposition	88
4.5	BED FORMS	90
4.5.1	The formation of current-produced bed forms	90
4.5.2	Current flow and bed forms	90
4.6	SUMMARY OF CHAPTER 4	92

CHAPTER 5 — BEACHES AND THE LITTORAL ZONE

5.1	THE DIVISIONS OF THE LITTORAL ZONE	95
5.1.1	Zones of wave action	96
5.1.2	The sediment profile	96
5.2	SEDIMENT MOVEMENT IN THE LITTORAL ZONE	98
5.2.1	Orbital velocities of waves and bed shear stess	98
5.2.2	Onshore and offshore movement of sediment by waves	99
5.2.3	The longshore transport of sediment by wave-generated currents	101
5.2.4	The longshore sediment transport rate	104
5.2.5	Sediment transport in combined waves and currents	105
5.3	BEACH PROFILES	106
5.3.1	Beach profile and grain size	106
5.3.2	Beach profile and wave type	106
5.4	BEACH MATERIALS AND SEDIMENTARY STRUCTURES	107
5.4.1	Beach materials	108
5.4.2	Sedimentary structures	108
5.5	SUMMARY OF CHAPTER 5	109

CHAPTER 6	TIDAL FLATS AND ESTUARIES	
6.1	**SEDIMENT TRANSPORT AND DEPOSITION ON TIDAL FLATS**	112
6.1.1	Sediment distribution on tidal flats	112
6.1.2	Low-latitude tidal flats	115
6.2	**ESTUARIES**	116
6.2.1	Estuarine types	116
6.2.2	Sedimentation in estuaries	122
6.2.3	Sedimentation in different types of estuaries	123
6.2.4	Preservation of the dynamic balance of estuaries	125
6.3	**SUMMARY OF CHAPTER 6**	126

CHAPTER 7	DELTAS	
7.1	**THE STRUCTURE OF A DELTA**	130
7.2	**MIXING AND SEDIMENT DEPOSITION AT DISTRIBUTARY MOUTHS**	132
7.2.1	River-dominated deltas	132
7.2.2	Tide-dominated deltas	137
7.2.3	Wave-dominated deltas	139
7.2.4	Other types of delta	140
7.3	**SUMMARY OF CHAPTER 7**	141

CHAPTER 8	SHELF SEAS	
8.1	**SHELF SEDIMENTS**	144
8.2	**SHELF PROCESSES**	146
8.2.1	Ocean currents	146
8.2.2	Local wind-driven currents	146
8.2.3	Tidal currents	148
8.2.4	Storm surges	152
8.2.5	Wave action	152
8.3	**BED FORMS ON THE CONTINENTAL SHELF**	153
8.3.1	Scour hollows, furrows and sandribbons	154
8.3.2	Sandwaves and other transverse bed forms	155
8.3.3	Sand-banks	156
8.4	**SHELF PROCESSES AND SEA-BED RESOURCES**	156
8.4.1	Aggregates	157
8.4.2	Placer deposits	158
8.4.3	Phosphorites	159
8.5	**SUMMARY OF CHAPTER 8**	159

SUGGESTED FURTHER READING	161
ANSWERS AND COMMENTS TO QUESTIONS	162
INDEX	187

ABOUT THIS VOLUME

This is one of a series of Volumes on Oceanography. It is designed so that it can be read on its own, like any other textbook, or studied as part of S330 *Oceanography*, a third level course for Open University students. The science of oceanography as a whole is multidisciplinary. However, different aspects fall naturally within the scope of one or other of the major 'traditional' disciplines. Thus, you will get the most out of this Volume if you have some previous experience of studying physics and/or geology. Other Volumes in this Series lie more within the fields of chemistry or biology (and their associated sub-branches) according to subject matter.

Chapter 1 describes the qualitative aspects of water waves, briefly reviews modern methods of wave measurement, and explores some of the simple relationships of wave dimensions and characteristics. It also examines the concept of wave energy, the behaviour of waves as they approach the shore and expend that energy in breaking, and the features and causes of unusual waves.

Tides are a special type of wave, and Chapter 2 outlines the mechanism whereby tides are generated by the gravitational attractions of the Sun and Moon, but constrained by the configuration of the ocean basins. The effects of tides are most evident in shallow seas and on the shoreline, and this Chapter also deals with the interaction of the tide with shoals, coasts and estuaries, and with the prediction of both normal and abnormal tides.

In coastal and shallow marine areas, waves and tidal currents are responsible for sediment movement and deposition. Chapter 3 introduces the nature of shallow marine sediments and the types of environments in which they are deposited. Chapter 4 goes on to consider, in general terms, the physical conditions that lead to the erosion, transport and deposition of sediment by flowing water. Some of the problems of applying the theory of fluid flow to the natural marine environment are discussed. The theory is somewhat complex and so a greater number of questions are used to help you work through the text.

Chapter 5 examines the conditions under which sediment is moved by waves, the rate at which it is moved and the way in which waves enhance currents. The reasons why some beaches have steep slopes and some have gentle slopes are also discussed.

Chapter 6 examines two types of coastal areas where tidal processes are more important than wave processes: tidal flats and estuaries. The pattern of sedimentation on tidal flats is controlled by the flood and ebb of the tides. Estuaries vary considerably in character due to variations in tidal range and river discharge which affect patterns of water circulation and sedimentation, and the extent to which seawater and river water mix.

When the sediment discharge from a river is so high that waves and tidal currents are unable to disperse it at the river mouth, a delta accumulates seawards of the mouth. Chapter 7 explains how the differences in the relative influences of rivers, tidal currents and wave energy lead to differences in sediment dispersal and give various types of deltas their characteristic shapes. The estuarine mixing processes discussed in Chapter 6 are seen to apply to processes at distributary mouths.

Under certain circumstances, sea-floor sediments can be moved in the deeper waters of the offshore zone. Chapter 8 outlines how currents and waves can affect sediments in water as deep as the shelf break, and considers how sediment transport paths across the sea-bed in current-dominated shelf seas can be determined. Finally, an outline is given of the mineral resources of continental shelf areas.

You will find that the terms 'speed' and 'velocity' are used frequently throughout. Strictly speaking, *speed* is the rate at which a particular distance is covered and the units are metres per second (ms^{-1}). *Velocity* is a quantity that specifies both speed and the direction of motion, and its units are also ms^{-1}. We have attempted to maintain this distinction in this Volume. For example, when discussing current flow in general terms, *current speed* is used; but *current velocity* is used instead when discussing the speed of a particular tidal current with a known direction. However, we have sometimes used velocity rather than speed where this is consistent with conventional usage, e.g. in the discussion of velocity profiles in Chapter 4 and water particle orbital velocities beneath waves in Chapter 5.

Finally, you will find questions designed to help you to develop arguments and/or test your own understanding as you read, with answers provided at the back of this Volume. Important technical terms are printed in **bold** type where they are first introduced or defined.

Notation used in this Volume

a	wave amplitude	q_l	longshore rate of sediment movement
c	wave speed		
c_g	wave group speed	q_s	rate of suspended load transport
C_b	concentration of suspended sediment close to the bed	R_d	rate of deposition of suspended load
d	water depth		
D	grain diameter	T	wave period
D_0	orbital velocity of a wave	u	horizontal component of flow velocity in net flow direction
E	wave energy		
f	wave frequency	$\bar{u}$	time-averaged mean horizontal velocity
F_g	gravitational force between two bodies	u_m	maximum horizontal orbital velocity
g	acceleration due to gravity	u_t	threshold orbital velocity
G	universal gravitational constant	u_*	shear velocity
		u_{*c}	critical shear velocity
H	wave height	v	horizontal component of flow velocity at right angles to net flow direction; also, linear velocities of the Earth's surface with respect to the Moon.
H_{max}	maximum wave height		
$H_{1/3}$	significant wave height		
k	wave number		
L	wavelength		
P	wave power	w	vertical component of flow velocity
P_l	longshore wave power		
q_b	rate of bedload transport	w_s	settling velocity
		z	height above a bed

6

z_0	roughness length	τ	shear stress
η	wave displacement (this symbol is also used for eddy viscosity)	τ_c	critical shear stress (for sediment erosion)
κ	von Karman constant	τ_{cu}	current-related shear stress
μ	molecular viscosity	τ_d	critical depositional shear stress
ρ	density	τ_{wc}	shear stress due to waves and currents
ρ_s	density of a sediment particle		
σ	angular frequency	τ_0	bed shear stress

CHAPTER 1 | WAVES

'…the chidden billow seems to pelt the clouds…'
Othello, Act II, Sc. I.

Sea waves have attracted attention and comment throughout recorded history. Aristotle (384–322 BC) observed the existence of a relationship between wind and waves, and the nature of this relationship has been the subject of study ever since. However, at the present day, understanding of the mechanism of wave formation and the way that waves travel across the oceans is by no means complete. This is partly because observations of wave characteristics at sea are difficult, and partly because mathematical models of wave behaviour are based upon the dynamics of idealized fluids, and ocean waters do not conform precisely with those ideals. Nevertheless, some facts about waves are well established, at least to a first approximation, and the purpose of this Chapter is to outline the qualitative aspects of water waves, and to explore some of the simple relationships of wave dimensions and characteristics.

We start by examining the dimensions of an idealized water wave, and the terminology used to describe them (Figure 1.1).

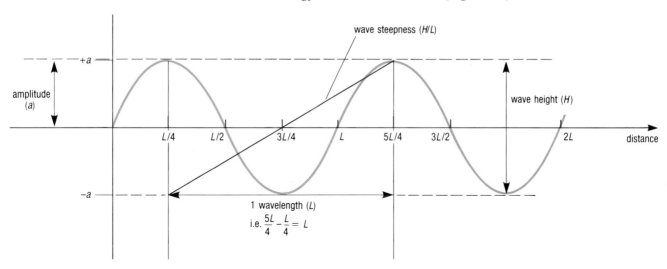

Figure 1.1 Vertical profile of two successive idealized ocean waves, showing their linear dimensions and sinusoidal shape.

Wave height (H) refers to the overall vertical change in height between the wave crest (or peak) and the wave trough. The wave height is twice the wave **amplitude** (a). **Wavelength** (L) is the distance between two successive peaks (or two successive troughs). **Steepness** is defined as wave height divided by wave length (H/L), and, as can be seen in Figure 1.1, is not the same thing as the slope between a wave crest and its adjacent trough.

In addition to having dimensions in space at a fixed instant in time (Figure 1.1), waves have dimensions in time at a fixed point in space. The time interval between two successive peaks (or two successive troughs) passing a fixed point is known as the **period** (T), and is measured in seconds. The number of peaks (or the number of troughs) which pass a fixed point per second is known as the **frequency** (f).

QUESTION 1.1   If a wave has a frequency of $0.2\,\mathrm{s}^{-1}$, what is its period?

As the answer to Question 1.1 shows, period is the reciprocal of frequency. We will return to this concept in Section 1.2.

1.1   WHAT ARE WAVES?

Waves are a common occurrence in everyday life, and include such examples as sound, the motion of a plucked guitar string, ripples on a pond, and the billows on the ocean. It is not easy to define a wave. Before attempting to do so, let us consider some of the characteristics of wave motion:

1   A wave transfers a disturbance from one part of a material to another. (The disturbance caused by dropping a stone into a pond is transmitted across the pond by ripples.)
2   The disturbance is propagated through the material without any substantial overall motion of the material itself. (A floating cork merely bobs up and down on the ripples, but experiences very little overall movement in the direction of travel of the ripples.)
3   The disturbance is propagated without any significant distortion of the wave form. (A ripple shows very little change in shape as it travels across a pond.)
4   The disturbance appears to be propagated with constant speed.

If the material itself is not being transported by wave propagation, then what *is* being transported?

The answer, 'energy', provides a reasonable working definition of wave motion—a process whereby energy is transported across or through a material without any significant overall transport of the material itself.

So, if energy, and not material, is being transported, what is the nature of the movement observed when ripples cross a pond?

There are two aspects to be considered: first, the progress of the waves (which we have already noted), and secondly, the movement of the water particles themselves. Superficial observation of the effect of ripples on a floating cork suggests that the water particles move 'up and down', but closer observation will reveal that, provided the water is very much deeper than the ripple height, the cork is describing a nearly circular path in a vertical plane, parallel with the direction of wave movement. In a more general sense, the particles are displaced from an equilibrium position, and then return to that position. Thus, the particles experience a displacing force and a restoring force. The nature of these forces is often used in the descriptions of various types of waves.

1.1.1   TYPES OF WAVES

All waves can be regarded as **progressive waves**, in that energy is moving through, or across the surface of, the material.

The so-called **standing wave**, of which the plucked guitar string is an example, can be considered as the sum of two progressive waves of equal dimensions, but travelling in opposite directions. We examine this in more detail in Section 1.5.4.

Waves which travel through the material are called body waves. Examples of body waves are seismic P- and S-waves, and sound waves, but our main concern in this Volume is with **surface waves**. The most familiar surface waves are those which occur at the interface between atmosphere and ocean, but surface waves can occur at the interface between any two bodies of fluid. For example, waves can occur at an interface between two layers of ocean water of differing densities. Because the interface is a surface, such waves are, strictly speaking, surface waves, but oceanographers usually refer to them as **internal waves**. Oscillations are more easily set up at an internal interface than at the sea-surface, because the difference in density between two water layers is smaller than that between water and air. Hence, less energy is required to generate internal waves than surface waves of similar amplitude. Internal waves travel more slowly than surface waves of comparable amplitude, and are of importance in the context of vertical mixing processes in the oceans. Surface waves are caused either by forces resulting from relative motion between two layers of fluid, as for example the wind blowing over the sea, or by an external force that disturbs the fluid. Examples of such external forces range from raindrops falling into a pond, through diving gannets, ocean-going liners and earthquakes, to the gravitational attractions of the Sun and Moon.

Waves that are caused by periodic forces, such as the effect of the Sun and Moon causing the tides, will have periods coinciding with the causative forces. This aspect is considered in more detail in Chapter 2. Most other waves, however, result from a non-periodic disturbance of the water. The water particles are displaced from an equilibrium position, and to regain that position require a restoring force. In the case of water waves, the particle motion resulting from the restoring force acting upon one wave cycle provides the displacing force acting upon the next cycle. Such alternate displacements and restorations establish a characteristic oscillatory 'wave motion', which in its simplest form has sinusoidal characteristics (Figures 1.1 and 1.6), and is sometimes referred to as simple harmonic motion. In the case of surface waves on water, there are two such restoring forces which maintain wave progress:

1 The gravitational force exerted by the Earth.
2 Surface tension, which is the tendency of water molecules to stick together and present the smallest possible surface to the air. So far as the effect on water waves is concerned, it is as if a weak elastic skin were stretched over the water surface.

Water waves are affected by both of these forces. In the case of waves with wavelengths less than about 1.7 cm, the principal maintaining force is surface tension, and such waves are known as **capillary waves**. Capillary waves are important in the context of remote sensing of the oceans (Section 1.6.1). However, the main interest of oceanographers lies with surface waves of wavelengths greater than 1.7 cm, and the principal maintaining force for such waves is gravity; hence they are known as **gravity waves**. Figure 1.2 summarizes some wave types and their causes.

Not all waves are displaced in a vertical plane. Because atmosphere and oceans are on a rotating Earth, variation of planetary vorticity with latitude causes deflection of atmospheric and oceanic currents, and

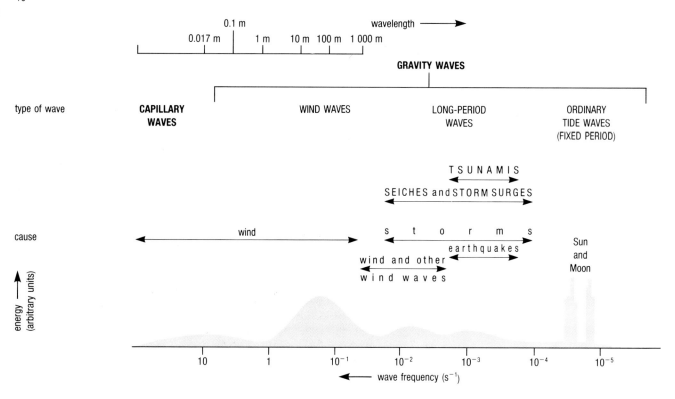

Figure 1.2 Types of surface waves, showing the relationships between wavelength, wave frequency, the nature of the displacing forces, and the relative amounts of energy in each type of wave. Unfamiliar terms will be explained later.

provides restoring forces which establish oscillations in a horizontal plane, so that easterly or westerly currents tend to swing back and forth about an equilibrium latitude. These large-scale waves are known as planetary or **Rossby waves**, and may occur as surface or as internal waves.

1.1.2 WIND-GENERATED WAVES ON THE OCEAN

In 1774, Benjamin Franklin said: 'Air in motion, which is wind, in passing over the smooth surface of the water, may rub, as it were, upon that surface, and raise it into wrinkles, which, if the wind continues, are the elements of future waves'.

In other words, if two fluid layers having differing speeds are in contact, and there is frictional stress between them, there is a transfer of energy. At the sea-surface, most of the transferred energy results in waves, although a small proportion is manifest as wind-driven currents. In 1925, Harold Jeffreys suggested that waves obtained energy from the wind by virtue of pressure differences caused by the sheltering effect provided by wave crests (Figure 1.3).

Although Jeffreys' hypothesis fails to explain the formation of very small waves, it does seem to work if:

1 Wind speed exceeds wave speed.
2 Wind speed exceeds $1\,\mathrm{m\,s^{-1}}$.
3 The waves are steep enough to provide a sheltering effect.

Empirically, it can be shown that the sheltering effect is at a maximum when wind speed is approximately three times the wave speed. In the open oceans, most wind-generated waves have a steepness (H/L) of about 0.03 to 0.06. In general, the greater the amount by which wind speed

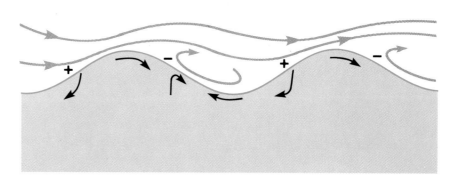

Figure 1.3 Jeffreys' 'sheltering' model of wave generation. Curved lines indicate air flow; short, straight arrows show water movement, which will be explained more fully in Section 1.2.1. The rear face of the wave against which the wind blows experiences a higher pressure than the front face, which is sheltered from the force of the wind. Air eddies are formed in front of each wave, leading to differences in air pressure. The excesses and deficiencies of pressure are shown by plus and minus signs respectively. The pressure difference pushes the wave along.

exceeds wave speed, the steeper the wave. However, as we shall see later, wave speed in deep water is not related to wave steepness, but to wavelength—the greater the wavelength, the faster the wave travels.

QUESTION 1.2 Two waves have the same height, but differing steepness. Which of the two waves will travel the faster?

Consider the sequence of events if, after a period of calm weather, a wind starts to blow, rapidly increases to a gale, and continues to blow at constant gale force for some considerable time. No significant wave growth occurs until the wind speed exceeds $1\,\mathrm{m\,s^{-1}}$. Then, small steep waves form as the wind speed increases. Even after the wind has reached a constant gale force, waves continue to grow with increasing rapidity until they reach a size and wavelength (and hence a speed) which corresponds to one-third of the wind speed. Beyond this point, the waves continue to grow in size, wavelength and speed, but at an ever-diminishing rate. On the face of it, one might expect that wave growth would continue until wave speed was the same as wind speed. However, in practice wave growth ceases whilst wave speed is still at some value below wind speed. This is because:

1 Some of the wind energy is transferred to the ocean surface via a tangential force, and thus produces a surface current.

2 Some wind energy is dissipated by friction.

3 Energy is lost from larger waves as a result of **white-capping**, i.e. breaking of the tip of the wave crest because it is being driven forward by the wind faster than the wave itself is travelling. Much of the energy dissipated during white-capping is converted into forward momentum of the water itself, reinforcing the surface current initiated by process 1 above.

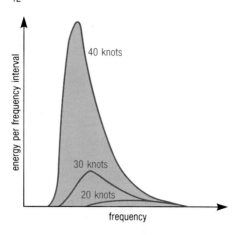

Figure 1.4 Wave energy spectra for three fully developed seas, related to wind speeds of 20, 30 and 40 knots (10.3, 15.45 and 20.6 ms⁻¹ respectively). The area under each curve is a measure of the total energy in that particular wave field.

1.1.3 THE FULLY DEVELOPED SEA

We have already seen that the size of waves in deep water is governed not only by the actual wind speed, but also by the length of time the wind has been blowing at that speed. Wave size also depends upon the unobstructed distance of sea, known as the **fetch**, over which the wind blows.

Provided the fetch is extensive enough, and the wind blows at constant speed for long enough, an equilibrium is eventually reached, in which energy is being dissipated by the waves at the same rate as the waves receive energy from the wind. Such an equilibrium results in a sea state called a **fully developed sea**, in which the size and characteristics of the waves are not changing. However, the wind speed is usually variable, so the ideal fully developed sea, with waves of uniform size, rarely occurs. Variation in wind speed produces variation in wave size, so, in practice, a fully developed sea consists of a range of wave sizes known as a **wave field**. Note that a range of wave sizes will also result from waves coming into the area from elsewhere, and from wave–wave interaction—a concept explained in Section 1.4.3. Oceanographers find it convenient to express a wave field as a spectrum of wave energies (Figure 1.4). The energy contained in a wave is proportional to the square of the wave height (see Section 1.4).

QUESTION 1.3 Examine Figure 1.4. Does the energy contained in a wave field increase or decrease as the average frequency of the constituent waves increases?

1.1.4 WAVE HEIGHT AND WAVE STEEPNESS

As was hinted in Section 1.1.3, the height of any particular wave is influenced by many wave components, each of different frequency and amplitude, which move into and out of phase with and across each other. In theory, if the heights and frequencies of all the contributing waves are known, it is possible to predict the heights and frequencies of the largest waves accurately. In practice, this is rarely possible. Figure 1.5 illustrates the range of wave heights which occur over a short time at one location—there is no obvious pattern to the variation of wave height.

Figure 1.5 A typical wave record, i.e. a record of variation in water level with time at one position.

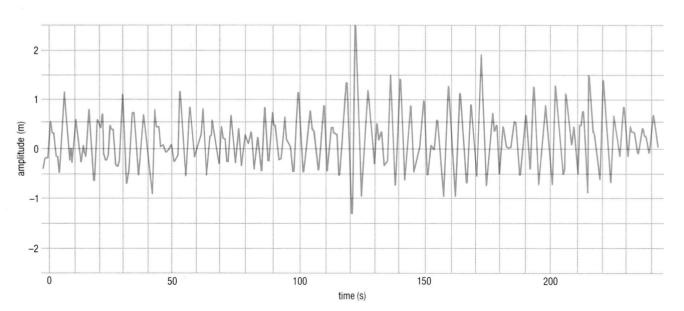

For many applications of wave research, it is necessary to choose a single wave height which characterizes a particular sea state. The one used most by oceanographers is the **significant wave height** or $H_{1/3}$, which is the average height of the highest one-third of all waves occurring in a particular time period. In any wave record, there will also be a maximum wave height, H_{max}. Prediction of H_{max} for a given period of time has great value in the design of structures such as flood barriers, harbour installations, and drilling platforms. To build these structures with too great a margin of safety would be unnecessarily expensive, but to underestimate H_{max} could have tragic consequences. However, it is necessary to emphasize the essentially random nature of H_{max}. The wave $H_{max(25\ years)}$ will occur on average once every 25 years. This does not mean such a wave will automatically occur every 25 years—there may be periods much longer than that without one. On the other hand, two such waves might appear next week.

As wind speed increases, so $H_{1/3}$ in the fully developed sea increases. The relationship between sea state, $H_{1/3}$ and wind speed is expressed by the **Beaufort Scale** (Table 1.1). The Beaufort Scale can be used to estimate wind speed at sea, but is valid only for waves generated within the local weather system, and assumes that there has been sufficient time for a fully developed sea to have become established.

Table 1.1 A selection of information from the Beaufort Wind Scale.

Beaufort No.	Name	Wind speed		State of the sea-surface	Wave height* (m)
		knots	ms⁻¹		
0	Calm	<1	0.0–0.2	Sea like a mirror.	0
1	Light air	1–3	0.3–1.5	Ripples with appearance of scales; no foam crests.	0.1–0.2
2	Light breeze	4–6	1.6–3.3	Small wavelets; crests have glassy appearance but do not break.	0.3–0.5
3	Gentle breeze	7–10	3.4–5.4	Large wavelets; crests begin to break; scattered white horses.	0.6–1.0
4	Moderate breeze	11–16	5.5–7.9	Small waves, becoming longer; fairly frequent white horses.	1.5
5	Fresh breeze	17–21	8.0–10.7	Moderate waves taking longer form; many white horses and chance of some spray.	2.0
6	Strong breeze	22–27	10.8–13.8	Large waves forming; white foam crests extensive everywhere and spray probable.	3.5
7	Moderate gale	28–33	13.9–17.1	Sea heaps up and white foam from breaking waves begins to be blown in streaks; spindrift begins to be seen.	5.0
8	Fresh gale	34–40	17.2–20.7	Moderately high waves of greater length; edges of crests break into sprindrift; foam is blown in well-marked streaks.	7.5
9	Strong gale	41–47	20.8–24.4	High waves; dense streaks of foam; sea begins to roll; spray may affect visibility.	9.5
10	Whole gale	48–55	24.5–28.4	Very high waves with overhanging crests; sea-surface takes on white appearance as foam in great patches is blown in very dense streaks; rolling of sea is heavy and visibility reduced.	12.0
11	Storm	56–64	28.5–32.7	Exceptionally high waves; sea covered with long white patches of foam; small and medium-sized ships might be lost to view behind waves for long times; visibility further reduced.	15.0
12	Hurricane	>64	>32.7	Air filled with foam and spray; sea completely white with driving spray; visibility greatly reduced.	>15

*$H_{1/3}$, i.e. the significant wave height.

14

The absolute height of a wave is less important to sailors than is its steepness (H/L). As was mentioned in Section 1.1.2, most wind-generated waves have a steepness in the order of 0.03 to 0.06. Waves steeper than this can present problems to shipping, but fortunately it is very rare for wave steepness to exceed 0.1. In general, wave steepness diminishes with increasing wavelength. The short choppy seas rapidly generated by local squalls are particularly unpleasant to small boats because the waves are steep, even though not particularly high. On the open ocean, very high waves can usually be ridden with little discomfort because of their relatively large wavelengths.

1.2 WAVE-FORMS

In order to simplify the theory of surface waves, we assume here that the wave-form is sinusoidal and can be represented by the curves shown in Figures 1.1 and 1.6. This assumption allows us to consider wave **displacement** (η) as simple harmonic motion, i.e. a cyclical variation in water level caused by the wave's passage. Figure 1.1 shows how the displacement varies over distance at a fixed instant in time—a 'snapshot' of the passing waves, whereas Figure 1.6 shows how wave displacement varies with time at a fixed point.

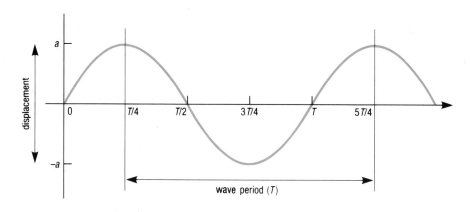

Figure 1.6   The displacement of an idealized wave at a fixed point, plotted against time.

Before examining displacement, let us remind ourselves of the relationship between period and frequency.

QUESTION 1.4   If sixteen successive wave troughs pass a fixed point during a time interval of one minute and four seconds, what is the frequency of the waves?

The displacement (η) of a wave at a fixed instant in time, or at a fixed point in space, varies between $+a$ (at the peak) and $-a$ (in the trough). Displacement is zero where $L=0$ on Figure 1.1 (and at intervals of $L/2$ from $L=0$ along the distance axis). Displacement is also zero at $T=0$ on Figure 1.6 (and at intervals of $T/2$ from $T=0$ along the time axis).

QUESTION 1.5 The peak, or crest, of a wave having a wavelength of 624m, a frequency of $0.05\,\mathrm{s}^{-1}$, and travelling in deep water, passes a fixed point P. What is the displacement (in terms of 'a') at P:

(a) 30 seconds after the peak has passed?

(b) 80 seconds after the peak has passed?

(c) 85 seconds after the peak has passed?

What is the displacement at a second point, Q, which is 312m away from P in the direction of wave propagation:

(d) when the displacement at P is zero?

(e) when the displacement at P is $-a$?

(f) five seconds after a trough has passed P?

The curves shown in Figures 1.1 and 1.6 are both sinusoidal. However, most wind-generated waves do not have simple sinusoidal forms. The steeper the wave, the further it departs from a simple sine curve. Very steep waves resemble a trochoidal curve, which is illustrated in Figure 1.7.

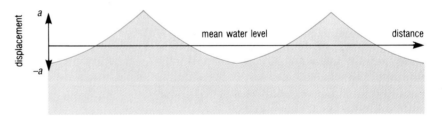

Figure 1.7   Vertical profile of two successive trochoidal waves.

A point marked on the tread of a car tyre will appear to trace out a trochoidal curve as the car is driven past an observer. Invert that pattern and you have the profile of a trochoidal water wave. We do not need to delve into the mathematical complexities of trochoidal wave-forms here, because the sinusoidal model is sufficiently accurate for an introduction to oceanic waves.

1.2.1   MOTION OF WATER PARTICLES

Water particles in a wave over deep water move in an almost closed circular path. At wave crests, the particles are moving in the same direction as wave propagation, whereas in the troughs they are moving in the opposite direction. At the surface, the orbital diameter corresponds to wave height, but the diameters decrease exponentially with increasing depth, until at a depth roughly equal to half the wavelength, the orbital diameter is negligible, and there is virtually no displacement of the water particles (Figure 1.8(a)).

It is important to realize that the orbits are only approximately circular. There is a small net component of forward motion, particularly in waves of large amplitude, so that the orbits are not quite closed, and the water, whilst in the crests, moves slightly further forward than it moves backward whilst in the troughs. This small net forward displacement of water in the direction of wave travel is termed **wave drift** (see Figure 1.8(b)). In shallow water, where depth is less than half the wavelength, the orbits become progressively flattened with depth (Figure 1.8(c)). The significance of this will be seen in Section 1.4.5, and in the Chapters on sediment movement.

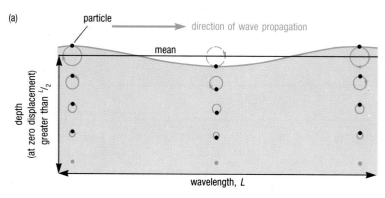

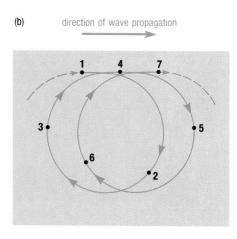

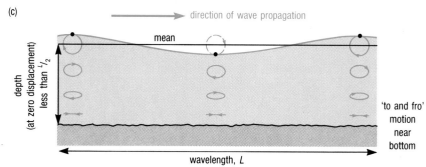

Figure 1.8(a) Particle motion in small deep-water waves, showing exponential decrease of the diameters of the orbital paths with depth.

(b) Particle motion in larger deep-water waves, showing drift.

(c) Particle motion in shallow-water waves, showing progressive flattening of the orbits near the sea-bed.

(d) Particle motions in internal waves. The orbits will only be truly circular if the layers are thick enough (i.e. greater than half the wavelength). The orbital diameters decrease with distance from the interface, as in the case of surface waves.

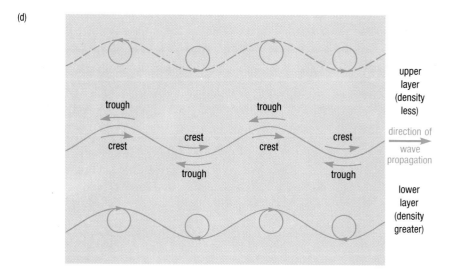

The nature of water particle motion in waves has some important practical applications. For example, a submarine only has to submerge about 150m to avoid the effects of even the most severe storm at sea, and knowledge of the exponential decrease of wave influence with depth has implications for the design of stable floating oil rigs.

The orbital motions relevant to internal waves are shown in Figure 1.8(d). Either side of the interface it can be seen that the water particles are moving in opposite directions, and under certain conditions may influence the movement of ships with a draught comparable with the depth to the interface.

1.2.2 SURFACE WAVE THEORY

As we have already hinted, there are mathematical relationships linking the characteristics of wavelength (L), wave period (T) and wave height (H) to wave speed in deep water and to wave energy.

First, let us consider **wave speed** (c) (the c stands for 'celerity of propagation').

Can you devise a simple formula for wave speed based on the symbols given above? Write it down before reading on.

The speed of a wave can be ascertained from the time taken for one wavelength to pass a fixed point,

$$\text{i.e.} \quad c = L/T \tag{1.1}$$

So, if we know any two of the variables in equation 1.1, we can calculate the third. However, we can do even better than that, as we shall see later, because there are other interrelationships between c, L and T.

Two terms you may meet in oceanographic literature are the **wave number** k, which is $2\pi/L$, and the **angular frequency** σ, which is $2\pi/T$.

QUESTION 1.6   How could c be expressed in terms of k and σ?

Note that the units of k are the number of waves per metre, and the units of σ are the number of cycles (waves) per second. From this information, the answer to Question 1.6 can be checked by going back to basic units, i.e. angular frequency/wave number

$$= \text{number of waves s}^{-1}/\text{number of waves m}^{-1}$$
$$= \text{wave speed (in ms}^{-1}\text{), i.e.}$$
$$c = \sigma/k \tag{1.2}$$

1.2.3 WAVE SPEED IN DEEP AND IN SHALLOW WATER

You may have noticed that, when wave speeds have been mentioned, we have been careful to state that the waves described were travelling over deep water. Thus you might have suspected that in shallow water, water depth has an effect on wave speed. If so, you were quite right. Wave speed can be represented by the equation:

$$c = \sqrt{\frac{gL}{2\pi} \tanh\left(\frac{2\pi d}{L}\right)} \tag{1.3}$$

where the acceleration due to gravity $g = 9.8 \text{ms}^{-2}$, $L =$ wavelength (m), and $d =$ water depth (m). Tanh is a mathematical function known as the hyperbolic tangent. All you need to know about it in this context is that if x is small, say less than 0.05, then $\tanh x \approx x$. If x is larger than π, then $\tanh x \approx 1$.

QUESTION 1.7   Armed with equation 1.3, and the information about the tanh function, work out the answers to the following questions:

(a)   What does equation 1.3 become if the water depth exceeds half the wavelength?

(b)   What does equation 1.3 become if the water depth is very much smaller than L?

Let us consider the implications of your answers to Question 1.7 in terms of factors affecting wave speed:

1   In water deeper than half the wavelength, the wavelength is the only variable which affects wave speed, and equation 1.3 approximates to:

$$c = \sqrt{\frac{gL}{2\pi}} \tag{1.4}$$

2   In water very much shallower than the wavelength (in practice when $d < L/20$), water depth is the only variable which affects wave speed, and equation 1.3 approximates to:

$$c = \sqrt{gd} \tag{1.5}$$

3   When d lies between $L/20$ and $L/2$, the full form of equation 1.3 is required. Hence, to calculate wave speed you would need to know wavelength and depth, and have access to a set of hyperbolic tangent tables, or a pocket calculator with hyperbolic functions on its keyboard.

In Section 1.2.2, mention was made of interrelationships between c, T and L. The answer to Question 1.7(a) (i.e. equation 1.4) allows us to explore these relationships. We saw, in equation 1.1, that $c = L/T$, so it is possible to combine equations 1.1 and 1.4.

QUESTION 1.8   Derive an equation for wavelength (L) in terms of period (T), using equations 1.1 and 1.4.

The answer to Question 1.8 provides an equation expressing L in terms of T, i.e.

$$L = \frac{gT^2}{2\pi} \tag{1.6}$$

A similar exercise, substituting the expression obtained for L from equation 1.6 into equation 1.1, will give c in terms of T. You may like to try this for yourself—you will need it for Question 1.10(a). Thus, it is possible, given only one of the wave characteristics c, T or L, to calculate either of the other two. Moreover, by working out the numerical values of the constants involved, the equations can be simplified.

QUESTION 1.9   Simplify equation 1.6 to give a numerical relationship between L and T^2.

QUESTION 1.10(a)   The period of a wave is 20 seconds. At what speed will it travel over deep water?

(b)   At what speed will a wave of wavelength 312m travel over deep water?

(c)   At what speeds will each of the waves referred to in (a) and (b) above travel in water of 12m depth?

The answer to Question 1.10(c) highlights an important conclusion about wave speed in shallow water. In water of a given depth, provided that depth is less than 1/20 of their wavelengths, all waves will travel at the same speed.

1.2.4 ASSUMPTIONS MADE IN SURFACE WAVE THEORY

The simple wave theory introduced in Sections 1.2.2 and 1.2.3 is a first-order approximation, and makes the following assumptions:

1 The wave shapes are sinusoidal.

2 The wave amplitudes are very small when compared with wavelengths and depths.

3 Viscosity and surface tension can be ignored.

4 Coriolis force and vorticity, both of which depend upon the Earth's rotation, can be ignored.

5 The depth is uniform, and the bottom has no bumps or hummocks.

6 The waves are not constrained or deflected by land masses, or by any other obstruction.

7 That three-dimensional waves behave in a way that is analogous to a two-dimensional model.

None of the above assumptions is valid in the strictest sense, but results predicted by using the simple model of surface wave behaviour are a close approximation to how wind-generated waves behave in practice.

1.3 WAVE DISPERSION AND GROUP SPEED

Those deep-water waves that have the greatest wavelengths and longest periods travel fastest, and thus are first to arrive in regions distant from the storm which generated them. This separation of waves by virtue of their differing rates of travel is known as **dispersion**, and equation 1.4 is sometimes known as the 'dispersion equation'.

The simple experiment of tossing a stone into a still pond shows that a band of ripples is created, which gets wider with increasing distance from the original disturbance. Ripples of greater wavelength progressively out-distance shorter ones—an example of dispersion in action. There is a second feature of the ripple band, which is not obvious at first sight. Each individual ripple travels faster than the band of ripples. A ripple appears at the back of the band, travels through it, and disappears out of the front. The speed of the band, called the **group speed**, is about half the wave speed of the individual ripples which travel through that band.

To understand the relationship between wave speed and group speed, the additive effect of two sets of waves (or wave trains) needs to be examined. If the difference between the wavelengths of two sets of waves is relatively small, the two sets will 'interfere' and produce a single set of resultant waves.

Figure 1.9 shows a simplified and idealized example of interference. Where the crests of the two wave trains coincide, the wave amplitudes are added, and the resultant wave has about twice the amplitude of the two original waves. Where the two wave trains are 'out of phase', i.e. where

(a)

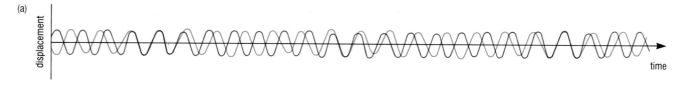

(b)

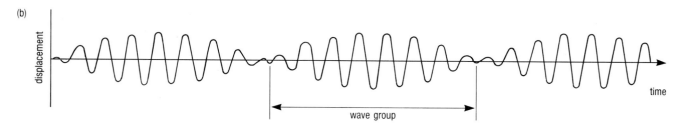

wave group

Figure 1.9(a) The merging of two wave trains (shown in red and blue) of slightly different wavelengths (but the same amplitudes), to form wave groups (b).

the crests of one wave train coincide with the troughs of the other, the amplitudes cancel out, and the water surface has minimal displacement.

The two component wave trains thus interact, each losing its individual identity, and combine to form a series of wave groups, separated by regions almost free from waves. The wave group advances more slowly than individual waves in the group, and thus in terms of the occurrence and propagation of waves, group speed is more significant than speeds of the individual waves in it. Individual waves do not persist for long in the open ocean, only as long as they take to pass through the group. Figure 1.10 shows the relationship between wave speed (sometimes called phase speed) and group speed in the open ocean.

If two sets of waves are interfering to produce a succession of wave groups, the group speed (c_g) is the difference between the two angular

Figure 1.10 The relationship between wave speed and group speed. As the wave advances from left to right, each wave moves through the group to die out at the front (e.g. wave 1), as new waves form at the rear (e.g. wave 6). In this process, the distance travelled by each individual wave as it travels from rear to front of the group is twice that travelled by the group as a whole. Hence, the wave speed is twice that of the group speed. Wave energy is contained within each group, and advances at the group speed.

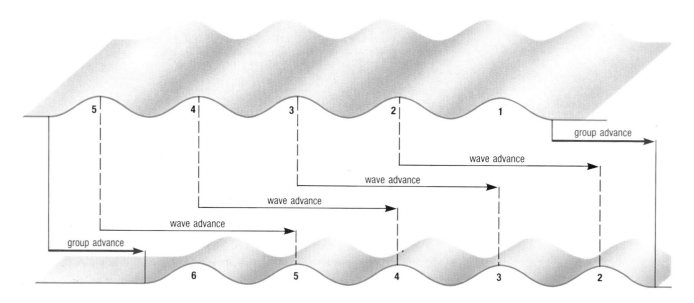

frequencies (σ_1 and σ_2) divided by the difference between the two wave numbers (k_1 and k_2 respectively), i.e.

$$c_g = \frac{\sigma_1 - \sigma_2}{k_1 - k_2} \tag{1.7}$$

In Section 1.2.3, it was shown that both T and L (and hence both σ and k) can be expressed in terms of c, the wave speed. If this is done for equation 1.7, c_g can be expressed in terms of the respective speeds, c_1 and c_2, of the two wave trains. The equation obtained is:

$$c_g = \frac{c_1 \times c_2}{c_1 + c_2} \tag{1.8}$$

If c_1 is nearly equal to c_2, then equation 1.8 simplifies to:

$$c_g \approx c^2/2c$$
or $c_g \approx c/2$ \hfill (1.9)

where c is the average speed of the two wave trains.

What happens to group speed when waves enter shallow water?

Equation 1.3 shows that as the water becomes shallower, wavelength becomes less important, and depth more important, in determining wave speed. As a result, wave speed in shoaling water becomes closer to group speed. Eventually, at depths less than $L/20$, all waves travel at the same depth-determined speed, there will be no wave–wave interference, and therefore in effect each wave will represent its own 'group'. Thus, in shallow water, group speed can be regarded as equal to wave speed.

1.4 WAVE ENERGY

The energy possessed by a wave is in two forms:

1   kinetic energy, which is the energy inherent in the orbital motion of the water particles; and

2   potential energy possessed by the particles when they are displaced from their mean position.

The total energy (E) per unit area of a wave is given by:

$$E = 1/8 \ (\rho g H^2) \tag{1.10}$$

where ρ is the density of the water (in $kg\,m^{-3}$), g is $9.8\,m\,s^{-2}$, and H is the wave height (m). The energy (E) is then in joules per square metre ($J\,m^{-2}$).

QUESTION 1.11   Would the total energy of a wave be doubled if its amplitude were doubled?

Equation 1.10 shows that wave energy is proportional to the square of the wave height.

1.4.1   PROPAGATION OF WAVE ENERGY

Figures 1.9 and 1.10 show that, in deep water, waves travel in groups, with areas of minimal disturbance between groups. Individual waves die out at the front of each group. It is obvious that no energy is being transmitted across regions where there are no waves, i.e. in between the

1

groups. Energy transmission is maximal where the waves in the group reach maximum size. It follows that the energy is contained within the wave group, and is propagated at the group speed. The rate at which energy is propagated per unit length of wave crest is called **wave power**, and is the product of group speed (c_g), and wave energy per unit area (E).

QUESTION 1.12(a) In the case of waves over deep water, what is the energy per square metre of a wave field made up of waves with an average amplitude of 1.3m? (Use $\rho = 1.03 \times 10^3 kg\,m^{-3}$.)

(b) What would be the wave power in kW per metre of crest length if the waves had a steepness of 0.04? (1 watt $= 1\,J\,s^{-1}$, and one kilowatt (kW) $= 10^3\,W$.)

Wave power has been seen by some as a possible source of pollution-free 'alternative energy', and has been used for some time on a small scale to recharge batteries on buoys carrying navigation lights. Harnessing wave energy on a large scale presents a number of problems.

1 Prevailing sea conditions must ensure a supply of waves with amplitudes sufficient to make conversion worthwhile.

2 Installations must not be a hazard to navigation, or to marine ecosystems. The nature of wave energy is such that rows of converters many kilometres in length are needed to generate amounts of electricity comparable with conventional power stations. These would form offshore barrages which might interfere with shipping routes, although sea conditions would be made calmer on the shoreward side. Calmer conditions, however, lead to reduced water circulation, less sediment transport, and increased growth of quiet-water plants and animals. Pollutants are less easily flushed away from such an environment.

3 The capital cost of such floating power stations and their related energy transmission and storage systems is enormous. Installations need to be robust enough to withstand storm conditions, yet sensitive enough to be able to generate power from a wide range of wave sizes. Such conditions are expensive to meet, and make it difficult for large-scale wave-energy schemes to be as cost effective as conventional energy sources. Relatively small-scale utilization of wave power is more feasible, as has been demonstrated by the Norwegians, who in 1985 brought into operation a wave-powered generator of 850kW. This machine was sited on the west coast of Norway, where waves funnelled into a narrow bay and increased substantially in height and thus in energy.

1.4.2   SWELL

The sea-surface is rarely still. Even when there is no wind, and the sea 'looks like a mirror', a careful observer will notice waves of very large wavelength (say 300 to 600m) and only a few centimetres amplitude. At other times, a sea may include locally generated waves of small wavelength, and travelling through these waves, possibly at a large angle to the wind, other waves of much greater wavelength. Such long waves are known as **swell**, which is simply defined as waves that have been generated elsewhere and have travelled far from their place of origin.

Systematic observations show that local winds and waves have very little effect on the size and progress of swell waves. Swell seems able to pass through locally generated seas without hindrance or interaction. This is because once swell waves have left the storm area, their wave height gradually diminishes, due to attenuation (Section 1.4.3). Once wave height has diminished to a few centimetres, swell waves are not steep enough to be significantly influenced by the wind.

In the ocean we find waves travelling in many directions, resulting in an apparently confused sea. To achieve a complete description of such a sea-surface, the amplitude, frequency, and direction of travel of each component would be needed. The energy distribution of the sea-surface could then be calculated, but, as you might imagine, such a complex process would require expensive equipment to measure the wave characteristics, and computer facilities to perform the necessary calculations.

One or more components of a confused sea may be long waves or swell resulting from distant storms. In practice, about 90% of the energy of the sea-surface propagates within an angle of 45° either side of the wind direction. Consequently, waves generated by a storm in a localized region of a large ocean radiate outwards as a segment of a circle (Figure 1.11). As the circumference of the circle increases, so energy per unit length of wave must decrease, so that the total energy of the segment remains the same.

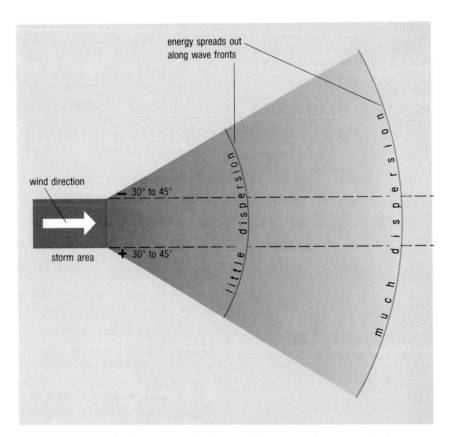

Figure 1.11 The spreading of swell from a storm centre, showing the area in which swell might be expected. As distance from the storm increases, the length of the wave crest increases, with a corresponding decrease in energy per unit length of wave. (By convention, direction is measured clockwise from North, hence the position of the plus and minus signs.)

The waves with the longest periods travel fastest, and progressively out-distance waves of higher frequencies (shorter periods). Near to the storm, dispersion is unlikely to be well defined, but the further one moves from the storm location, the more clearly separated waves of differing frequencies become.

QUESTION 1.13   Figure 1.12 shows two wave-energy spectra, (a) and (b) (*cf.* Figure 1.4). One represents the energy of the wave field in a storm-generating area; and the other represents the energy of the wave field in an area far away from the storm, but receiving swell from it. Which of the two profiles represents which situation?

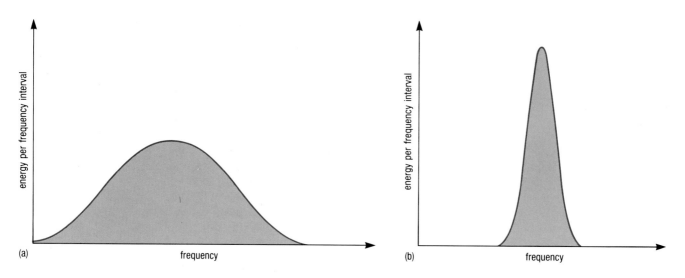

Figure 1.12   Wave energy spectra, each collected over a short time interval, for two areas (a) and (b) in the same ocean. One area is a storm centre, and the other is far away from the storm. (For use with Question 1.13.)

If you recorded the waves arriving from a storm a great distance (over 1000 km) away, you would, as time progressed, see the peak in the wave energy spectrum move progressively towards higher frequencies. By recording the frequencies of each of a series of swell waves arriving at a point, it would be possible to calculate each of their speeds. From the set of speeds a graph could be plotted to estimate the time and place of their origin. Before the days of meteorological satellites, this method was often used to pinpoint where and when storms had occurred in remote parts of the oceans.

1.4.3   ATTENUATION OF WAVE ENERGY

Attenuation involves loss or dissipation of wave energy, resulting in a reduction of wave height. Energy is dissipated in four main ways:

1   White-capping, which involves transfer of wave energy to the kinetic energy of moving water, thus reinforcing the wind-driven surface current (Section 1.1.2).

2   Viscous attenuation, which is only important for very high frequency capillary waves, and involves dissipation of energy into heat by friction between water molecules.

3   Air resistance, which applies to large steep waves soon after they have left the area in which they were generated and enter regions of calm or contrary winds.

4   **Non-linear wave–wave interaction**, which is more complicated than the simple (linear) combination of frequencies to produce wave groups as outlined in Section 1.3.

Non-linear interaction appears to be most important in the frequency range of 0.2 to $0.3 s^{-1}$. Groups of three or four frequencies can interact in complex non-linear ways, to transfer energy to waves of both higher and lower frequencies. A rough but useful analogy is that of the collision of two drops of water. A linear combination would simply involve the two drops coalescing into one big drop, whereas a non-linear combination is akin to a collision between the drops so that they split into a number of drops of differing sizes. The total amount of water in the drops (analogous to the total amount of energy in the waves) is the same before and after the collision.

Thus non-linear wave–wave interaction involves no loss of energy in itself, because energy is simply 'swapped' between different frequencies. However, the total amount of energy available for such 'swapping' will gradually decrease, because higher frequency waves are more likely to dissipate energy in the methods described under 1 and 2 above. For example, higher frequency waves are likely to be steep, and thus more prone to white-capping. As we saw in Section 1.4.2, in the case of established swell waves there is very little loss of wave energy apart from that caused by spreading over a progressively wider front (Figure 1.11). Wave attenuation is greatest in the storm-generating area, where there are waves of many frequencies, and hence more opportunities for energy exchange between waves of frequencies in the range 0.2 to $0.3s^{-1}$.

1.4.4 WAVE REFRACTION

Figure 1.13 shows an idealized linear wave crest (length s_1, between A and B) approaching a shoreline at an angle. Because the waves are travelling over shallow water, their speed is depth-determined (equation 1.5, $c = \sqrt{gd}$). Depth at A exceeds depth at B, hence the wave at A will travel faster than the wave at B. This will tend to 'swing' the wave crest to an alignment parallel to the depth contours; a phenomenon known as **refraction**.

Can the extent of refraction be quantified?

Refraction of waves in progressively shallowing water can be described by a relationship similar to Snell's law, which describes refraction of light rays through materials of different refractive indices.

Rays can be drawn perpendicular to the wave crests, and will indicate the direction of wave movement. The angles between these wave rays and lines drawn perpendicular to the depth contours can be related to wave speeds at various depths. In Figure 1.13, a wave ray approaching shoaling water at an angle θ_1, where water depth is d_1, will be at an angle θ_2 when it reaches depth d_2. Angles θ_1 and θ_2 are related to wave speed by:

$$\frac{\sin \theta_1}{\sin \theta_2} = \frac{c_1}{c_2} = \frac{\sqrt{gd_1}}{\sqrt{gd_2}} = \frac{\sqrt{d_1}}{\sqrt{d_2}} = \sqrt{\frac{d_1}{d_2}} \qquad (1.11)$$

where c_1 and c_2 are the respective wave speeds at depths d_1 and d_2.

You might ask: why go to the trouble of drawing perpendiculars? Why not simply use the angles between wave crests and bottom contours?

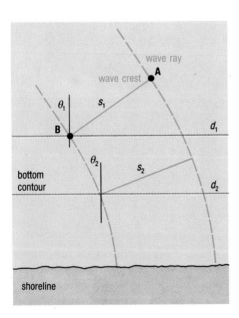

Figure 1.13 Plan view illustrating the relationship between wave approach angle (θ), water depth (d), and wave crest length (s). The wave rays (broken blue lines), are normal to the wave crests, and are the paths followed by points on the wave crests. For further explanation, see text.

Well, of course, one could do that, and obtain exactly the same relationships between the relevant angles, depths and wave speeds. However, wave rays are often more useful than wave crests in determining regions that are likely to experience high waves due to the effects of refraction.

26

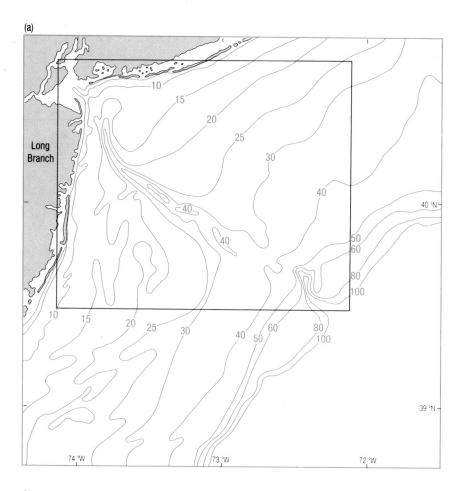

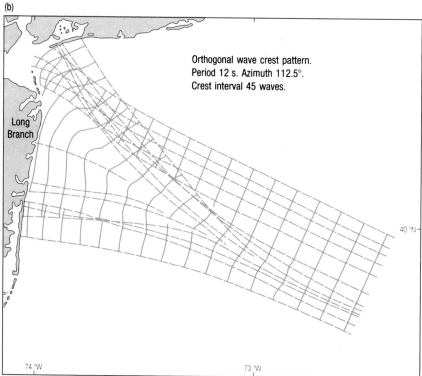

Orthogonal wave crest pattern.
Period 12 s. Azimuth 112.5°.
Crest interval 45 waves.

Figure 1.14(a) Bathymetric map of the continental shelf off New York harbour, at the mouth of the Hudson River, on the Atlantic coast of the USA. The rectangle shows the area of map (b). The position of the Hudson Canyon can be deduced from the submarine contours (in fathoms). (b) Orthogonal wave crest (wave front) pattern of part of Long Branch beach. The crest intervals are of 45 waves of period 12s.

Consider a length, s_1, of ideal wave crest, with energy per unit length E_1, which is bounded by two wave rays, as in Figure 1.13. To a first approximation, we may assume that the total energy of the wave crest between these two rays would remain constant as the wave progresses. Therefore, if the two rays converge, the same amount of energy is contained within a shorter length of wave crest, so that, for the total wave energy to remain constant, the wave would have to become higher (equation 1.10). If the wave rays were to diverge then the wave would become lower.

If the two wave rays, as they finally approach the shore, are separated by a length s_2, as in Figure 1.13, and if the wave energy is conserved, then the final wave energy must equal the initial wave energy, i.e. $E_1s_1 = E_2s_2$, or in terms of wave heights:

$$H_1^2 s_1 = H_2^2 s_2$$

(Remember $E = 1/8\ (\rho g H^2)$, eqn. 1.10.)

Note that for simplicity, s_2 in Figure 1.13 is the same length as s_1. However, it is common for wave rays to converge or diverge. Wave refraction diagrams can be plotted for a region by using the wave of most common period and the most common direction of approach, and areas in which wave rays are focused or defocused can be identified.

QUESTION 1.14 Figure 1.14 is a bathymetric map and storm wave refraction diagram for the Hudson River submarine canyon on the Atlantic coast of the USA. In what area covered by the refraction diagram would you advise fishermen to leave their boats with least likelihood of major damage, and why?

We can estimate increase or decrease in wave size by measuring the distances between wave rays, and applying equation 1.12. This method is quite useful provided wave rays neither approach each other too closely, nor cross over, as in these cases the waves become high, steep and unstable, and so simple wave theory becomes inadequate.

1.4.5   WAVES APPROACHING THE SHORE

As you have seen in the previous Section, refraction can change wave height, but it is also apparent that waves coming straight onto a beach increase in height and steepness until they break.

Figure 1.15(a) shows a length of wave crest, s, which is directly approaching a beach. As the water is shoaling, the wave crest passes a first point d_1, where the water is deeper than at a second point d_2, nearer the shore. We assume that the amount of energy within this length of wave crest remains constant, the wave is not yet ready to break, and that water depth is less than 1/20 of the wavelength (i.e. equation 1.5 applies: $c = \sqrt{gd}$). Because wave speed in shallow water is related to depth, the speed c_1 at depth d_1 is greater than the speed c_2 at depth d_2. If energy remains constant per unit length of wave crest, then

$$E_1 c_1 s = E_2 c_2 s$$

or $$\frac{E_2}{E_1} = \frac{c_1}{c_2} \tag{1.13}$$

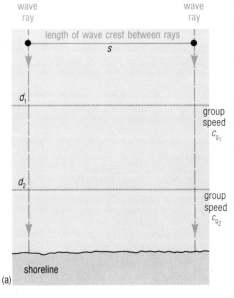

Figure 1.15(a) Plan view illustrating changes in energy as waves approach the shore. For explanation, see text.

and because energy is proportional to the square of the wave height (equation 1.10) then we can write

$$\frac{E_2}{E_1} = \frac{c_1}{c_2} = \frac{H_2^2}{H_1^2} \qquad (1.14)$$

Thus, both wave energy and the square of the wave height are inversely proportional to wave speed in shallow water.

This relationship is straightforward once the wave has entered shallow water. But what happens to energy in waves travelling at group speed over deep water as they move into shallow water where speed is wholly depth-determined?

This is quite a difficult question, best answered by considering the highly simplified case illustrated in Figure 1.15(b). Imagine waves travelling shoreward over deep water (depth greater than half the wavelength). Wave speed is then governed solely by wavelength (equation 1.4, $c = \sqrt{gL/2\pi}$). The energy is being propagated at the group speed (c_g) which is approximately half the wave speed (c). Once the waves have moved into shallow water, wave speed becomes governed solely by depth and is much reduced. Remember from Section 1.3 that group speed is equal to wave speed in shallow water. The rate at which energy arrives from offshore of point X on Figure 1.15(b) is equal to the rate at which

Figure 1.15(b)   Vertical profile of a highly simplified shoreline. The energy is being brought in from offshore at the same rate as it is being removed as the waves break. It follows that if c_{g1} is greater than c_{g2}, then there must be more energy per unit length of wave, and a greater wave height, in those waves travelling at c_{g2}.

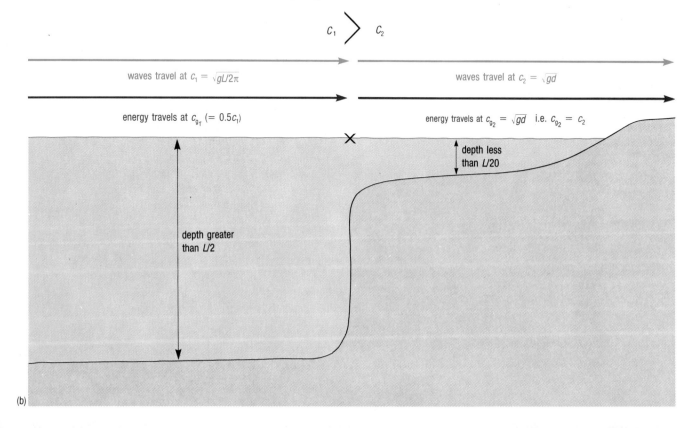

(b)

energy moves inshore of X, so if the group speed in shallow water is less than half the original wave speed (and hence less than the original group speed) in deep water, the waves will show corresponding increases in height and in energy. Of course, a 'real' coastline shows a much less abrupt change from deep to shallow water than does Figure 1.15(b), and calculations of intermediate values of group speeds, wave speeds, heights and energies become more complicated, because at depths between 1/2 and 1/20 of the wavelength, equation 1.3 must be used.

As waves move into shallow water, the circular orbits of the water particles become flattened (Figure 1.8(c)) and some wave energy will be used in moving sediments to and fro on the sea-bed. The shallower the slope of the immediate offshore region, the more energy will be lost from waves before they finally break.

1.4.6   WAVES BREAKING UPON THE SHORE

A breaking wave is a highly complex system. Even some distance before the wave breaks, its shape is substantially distorted from a simple sinusoidal wave. Hence the mathematical model of such a wave is more complicated than that we have assumed in this Chapter. As a wave breaks, the energy it received from the wind is dissipated. Some energy is reflected back out to sea, the amount depending upon the slope of the beach—the shallower the angle of the beach slope, the less energy is reflected. Most of the energy is dissipated as heat in the final small-scale mixing of foaming water, sand and shingle. Some energy is used in fracturing large rock or mineral particles into smaller ones, and yet more may be used to increase the height and hence the potential energy of the beach form. This last aspect depends upon the type of waves. Small gentle waves and swell tend to build up beaches, whereas storm waves tear them down.

Four major types of breaker are seen:

1   Spilling breakers are characterized by foam and turbulence at the wave crest. Spilling usually starts some distance from shore and is caused when a layer of water at the crest moves forward faster than the wave as a whole. Foam eventually covers the leading face of the wave. Such waves are characteristic of a gently sloping shoreline. A tidal bore (Section 2.4.3) is an extreme form of a spilling breaker. Breakers seen on beaches *during* a storm, when the waves are steep and short, are of the spilling type. They dissipate their energy gradually as the top of the wave spills down the front of the crest, which gives a violent and formidable aspect to the sea because of the more extended period of breaking.

2   Plunging breakers are the most spectacular type. The classical form, much beloved by surf-riders, is arched, with a convex back and a concave front. The crest curls over and plunges downwards with considerable force, dissipating its energy over a short distance. Plunging breakers on beaches of relatively gentle slope are usually associated with the long swells generated by distant storms. Locally generated storm waves seldom develop into plunging breakers on gently sloping beaches, but may do so on steeper ones.

3   Collapsing breakers are similar to plunging breakers, except that instead of the crest curling over, the front face collapses. Such breakers occur on beaches with moderately steep slopes, and under moderate wind conditions.

30

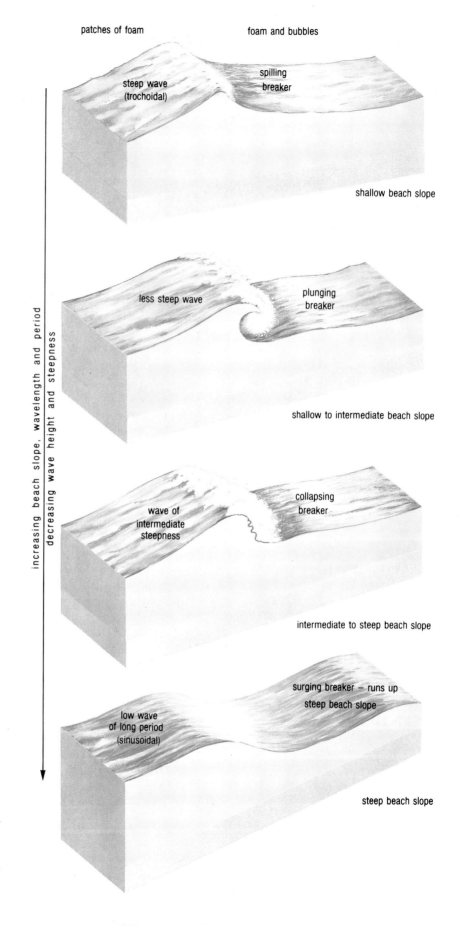

patches of foam   foam and bubbles

steep wave
(trochoidal)

spilling
breaker

shallow beach slope

less steep wave

plunging
breaker

shallow to intermediate beach slope

wave of
intermediate
steepness

collapsing
breaker

intermediate to steep beach slope

surging breaker – runs up
steep beach slope

low wave
of long period
(sinusoidal)

steep beach slope

increasing beach slope, wavelength and period

decreasing wave height and steepness

Figure 1.16   Four types of breaker and their
relationships to beach slope, wave period, length,
height and steepness.

4   Surging breakers are found on the very steepest beaches. Surging breakers are typically formed from long, low waves, and the front faces and crests remain relatively unbroken as the waves slide up the beach.

Figure 1.16 illustrates the relationship between wave steepness, beach steepness and breaker type.

QUESTION 1.15   If you observed plunging breakers on a beach and walked along towards a region where the beach became steeper, what different types of breaker might you expect to see?

From the description, Figure 1.16 and the answer to Question 1.15, it can be seen that the four types of breaker form a continuous series. The spilling breaker, characteristic of shallow beaches and steep waves (i.e. with short periods and large amplitudes), forms one end member. At the other end of the series is the surging breaker, characteristic of steep beaches and of waves with long periods and small amplitudes.

For a given beach, the arrival of waves steeper than usual will tend to give a type of breaker nearer the 'spilling' end of the series, whereas calmer weather favours the surging type.

1.5   WAVES OF UNUSUAL CHARACTER

Waves of unusual character may result from any one of a number of conditions, such as a particular combination of wave frequencies; the constraining effect of adjacent land masses; interaction between waves and ocean currents; or a submarine earthquake. The destructive effects of abnormally large waves are well known, and hence prediction of where and when they will occur is of extreme importance to all who live or work beside or upon the sea.

1.5.1   WAVES AND CURRENTS

Anyone who regularly sails a small boat into and out of estuaries will be well aware that at certain states of the tide the waves can become abnormally large and uncomfortable. Such large waves are usually associated with waves propagating against an ebbing tide. Because the strength of the tidal current varies with position as well as with time, waves propagating into an estuary during an ebb tide often advance into progressively stronger counter-currents.

Consider a simple system of deep-water waves, moving from a region with little or no current (A) into another region (B) where there is a current flowing parallel to the direction of wave propagation. Imagine two points, one in region A and one in region B, each of which are fixed with respect to the sea-bed. The number of waves passing each point in a given time must be the same, otherwise waves would either have to disappear, or be generated, between the two points. In other words, the wave period must be the same at each point.

How will wavelength and wave height be affected if the current is flowing (a) with or (b) against the direction of wave propagation?

Clearly, a current flowing with the waves will have the effect of increasing the speed of the waves, although the wave period (T) must remain constant, i.e.

$$T = \frac{L_0}{c_0} = \frac{L}{c + u} \qquad (1.15)$$

where L_0 = wavelength when current is zero;
c_0 = wave speed when current is zero;
L = wavelength in the current;
c = wave speed in the current;
u = speed of the current.

Because $c + u$ is greater than c_0, then for T to remain constant, L must be greater than L_0, i.e. the waves get longer. Moreover, the waves get correspondingly lower, because the rate of energy transfer depends upon group speed (half wave speed) and wave height. If the rate of energy transfer is to remain constant, then as speed increases, wave height must decrease. However, in practice, not all of the wave energy is retained in the wave system: some is transferred to the current, causing wave height to decrease still further.

Conversely, if the current flows counter to the direction of wave propagation, then L will decrease and the waves will get shorter and higher. Wave height will be further increased as a result of energy gained from the current. In theory, a point could be reached where wave speed is reduced to zero, so that a giant wave builds up to an infinite height (it can be shown mathematically that this occurs when the counter-current exceeds half the group speed of the waves in still water). However, in practice, as waves propagate against a counter-current of ever-increasing strength, the waves become shorter, steeper and higher until they become unstable and break, so that waves do not propagate against a counter-current of more than half their group speed.

Consider a wide estuary or fjord with a relatively narrow inlet from the sea. The only ocean waves that disturb this estuary during an ebbing tide are those that have speeds sufficiently high to overcome the effects of the counter-current. A current can also refract waves, i.e. change the direction of propagation. In such situations, the refraction diagram of the wave rays can be plotted in a similar way to that outlined in Section 1.4.4 (see also Section 1.5.2 and Figure 1.17(b)).

1.5.2 GIANT WAVES

The cultures of all seafaring nations abound with legends of ships being swamped by gigantic waves, and of sightings of waves of unbelievable size.

QUESTION 1.16   An elderly ex-seaman, in his cups, claims to have seen gigantic waves in the Southern Ocean, successive peaks of which took 30 seconds to pass, and which had wavelengths twice as long as his ship. Can you believe him?

Before dismissing the sailor's claim in Question 1.16 as a tall story, let us examine it more closely. Let us suppose his ship was travelling in the same direction as the waves and was being *overtaken* by them, and he had made the simple mistake of not taking account of the ship's velocity with respect to the waves when timing the intervals between successive peaks.

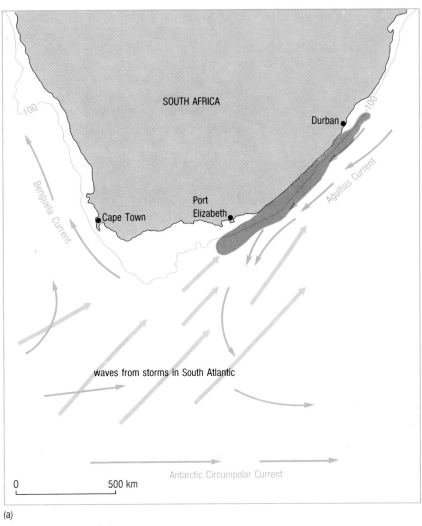

(a)

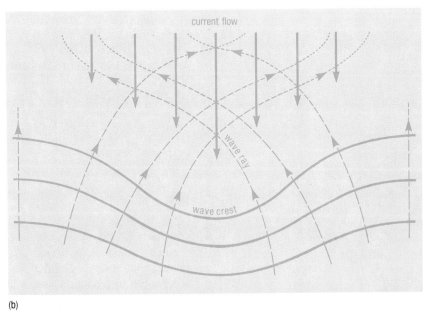

(b)

Figure 1.17(a) Giant waves have been recorded in the vicinity of the Agulhas Current (area shaded in red). Dark blue arrows show current direction, and light blue arrows indicate the direction of wave movement. The thin blue line is the 100 fathom contour.
(b) Diagram to illustrate how wave energy might be focused by lateral shear in a counter-current. The lengths of the dark blue arrows are proportional to the current velocity, curved solid lines are wave crests, and curved broken arrows are wave rays, with dashed extensions to indicate how waves could become trapped within the current if conditions were right.

An early objective method for estimating the height of large waves was to send a seaman to climb the rigging until he could just see the horizon over the top of the highest waves when the ship was in a wave trough. This technique was used by the *Venus* during her circumnavigation of the world from 1836 to 1839. She did not meet any particularly high waves during her voyage—the highest estimated by this method was about 8 m high, off Cape Horn. The highest reliably measured wave was that encountered by the US tanker *Ramapo*, en route from Manila to San Diego across the North Pacific in 1933. The ship was overtaken by waves having heights up to 34 m.

A region of the ocean which is infamous for encounters with giant waves is the Agulhas Current off the east coast of South Africa (Figure 1.17(a)). Waves travelling north-east from the southern Atlantic Ocean tend to be focused by the current (Figure 1.17(b)), and are further steepened and shortened by the counter-current effect outlined in Section 1.5.1. Wave periods of 14 seconds, with corresponding wavelengths of about 300 m, are quite common. Wave heights in this region can be of the order of 30 m which would result in a very steep wave (0.1). Such high and steep waves are sometimes preceded by correspondingly deep troughs, which are particularly dangerous as they can only be seen by vessels which are on the brink of the preceding wave.

Look again at Figure 1.5. The high wave occurring 122 seconds into the wave record is preceded by a particularly deep trough, but that is not the case for the high wave occurring at 173 seconds. As we saw in Section 1.1.4, ocean waves are rarely regular, and it is usually not possible to predict the heights of individual waves, nor the depths of individual troughs.

1.5.3   TSUNAMIS

Tsunami is a Japanese word for ocean waves of very great wavelength, caused either by a seismic disturbance, or by slumping of submarine sediment masses due to gravitational instability. Although commonly miscalled 'tidal waves', tsunamis are not caused by tidal influences. Tsunamis commonly have wavelengths of the order of hundreds of kilometres.

Although the tsunami travels at great speed in the open ocean, its wave height is small, usually in the order of one metre, and often remains undetected. As your answer to Question 1.18 indicates, even in the open ocean the ratio of wavelength to depth is such that a tsunami travels as a shallow-water wave, i.e. its speed is always governed by the depth of ocean over which it is passing. Thus, on reaching even shallower water, the speed diminishes, but the energy in the wave remains the same. Hence wave height must increase.

Great destruction can be wreaked by a tsunami. It is not unknown for people on board ships at anchor offshore to be unaware of a tsunami passing beneath them, but to witness the adjacent shoreline being pounded by large waves only a few seconds later. Tsunamis occur most frequently in the Pacific, because that ocean experiences frequent seismic activity. Accurate earthquake detection can give warning of the approach of tsunamis to coasts some distance from the earthquakes. Around and across the Pacific Ocean, a system of warning stations has long been established, of which Honolulu is the administrative and geographical centre.

1.5.4 SEICHES

A **seiche** is a standing wave, which can be considered as the sum of two progressive waves, travelling in opposite directions (Section 1.1). Seiches can occur in lakes, and in bays, estuaries or harbours which are open to the sea at one end. A seiche can be readily modelled by filling a domestic bath with water, and setting the water into oscillatory motion by moving a hand to and fro in the water. Figure 1.18(a) is an idealized vertical profile of a seiche. At either end of the container, water level is alternately high and low, whereas in the middle the water level remains constant. The length of the container (l) corresponds to half the wavelength (L) of the seiche.

Where the water level is constant (the **node**), the horizontal flow of water from one end of the container to the other is greatest. Where the fluctuation of water level is greatest (the **antinodes**), there is minimal horizontal movement of the water.

If the water depth divided by the length of the container is less than 0.1, then the waves can be considered to behave as shallow water waves, and the period of oscillation, T, is given by:

$$T = \frac{2l}{\sqrt{gd}} \tag{1.16}$$

where l = length of container; d = depth; and g = 9.8 ms^{-2}.

In most bays and estuaries, the water is relatively shallow compared with the seiche wavelength (L), and the period of the seiche is determined by the length of the basin and the depth of water in it.

In some basins, open to the sea at one end, it is possible for a node to occur at the entrance to the basin and an antinode at the landward end (Figure 1.18(b)). In this case, the length of the basin (l) corresponds to a quarter of the wavelength of the seiche (L). The corresponding equation for the period is therefore:

$$T = \frac{4l}{\sqrt{gd}} \tag{1.17}$$

T is also known as the **resonant period**. For standing waves to develop, the resonant period of the basin must be equal to the period of the wave motion or to a small whole number of multiples of that period.

QUESTION 1.19   A small harbour, open to the sea at one end, is 90m long and 10m deep at high water. What would be the effect of swell waves of period 18 seconds arriving at the harbour mouth?

Your answer to Question 1.19 is an example of how the arrival of waves of a certain frequency can create problems for moored vessels in small

36

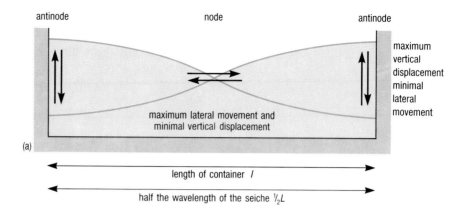

length of container *l*

half the wavelength of the seiche ½*L*

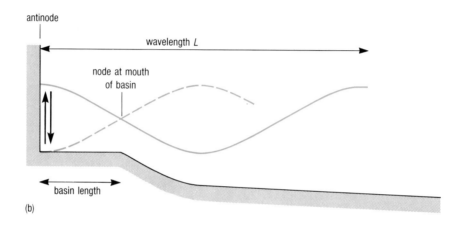

wavelength *L*

node at mouth of basin

basin length

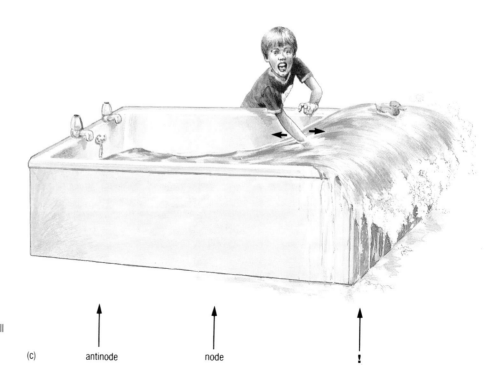

Figure 1.18(a) A simple standing wave.
(b) A quarter-wavelength standing wave in a small harbour.
(c) The hazards resulting from a seiche.

(c)   antinode   node   !

harbours by setting up a standing wave. Just as the seiche in a bath can be made to 'slop over' if you persist in wave generation for too long (Figure 1.18(c)), so a standing wave in a harbour may dash vessels against the harbour wall, or even throw them ashore. When the standing wave is at the low point of the antinode, there is also the danger of vessels being grounded, thus suffering damage to their hulls.

1.6 MEASUREMENT OF WAVES

Various instruments have been devised to measure wave characteristics. For example, a pressure gauge can be placed upon the sea-floor and will detect the frequency and size of waves passing above it. Pressure gauges with a sensitivity of about one part in one million are available. Such gauges can detect a water level change of less than a centimetre in a depth of several thousand metres of water. Another method is to place an instrument akin to an accelerometer in a moored buoy which then detects the rise and fall of waves displacing the buoy. Most wave measurement employs one or other of these methods.

1.6.1 SATELLITE OBSERVATIONS OF WAVES

The observation of ocean surface waves is particularly amenable to the use of satellite-based remote-sensing. Four main techniques are considered here, but the technical details are not important in the context of this Chapter:

1 Radar altimetry
A radar altimeter, mounted in a satellite, emits radar pulses at a rate of about 1000 pulses per second directly towards the sea-surface, and each reflected pulse is picked up by a sensor. Radar pulses pass through, and are not reflected by, water vapour in the atmosphere, but are reflected by the sea-surface. Two things are measured:

(i) The average time which the pulses take to travel to the sea-surface and back again, which enables the mean satellite-to-surface distance to be calculated. The accuracy with which the distance can be measured depends upon the sea state. For values of $H_{1/3}$ up to 8m, the accuracy is ± 8cm, and for values of $H_{1/3}$ over 8m, the accuracy is about $\pm x$cm, where x is the significant wave height in metres.

(ii) Changes in the shape and amplitude of the returning pulses, which can be used to give a measure of wave heights, from which the significant wave height ($H_{1/3}$) can be calculated. The accuracy achieved is about $\pm 10\%$ of $H_{1/3}$ for waves of significant wave height greater than 3m, and about ± 0.3m for smaller values of $H_{1/3}$.

QUESTION 1.20 Estimate the uncertainties which would be expected in (i) the mean satellite-to-surface distance, and (ii) the significant wave height, if they were measured by radar altimeter under the following conditions:

(a) Force 1 wind on the Beaufort Scale, no cloud;
(b) Force 4 wind on the Beaufort Scale, dense cloud cover;
(c) Force 10 wind on the Beaufort Scale, light cloud cover.

Use Table 1.1, and assume a steady sea state has been established in each case.

2 Synthetic aperture radar (SAR)

The procedure maps variations in the proportion of the radar signal that is back-scattered to the radar antenna by the sea-surface, and gives a measure of its 'roughness'. The technique involves the transmission of short radar pulses at an oblique angle to the sea-surface, and analysing the Doppler shift (i.e. apparent changes in signal frequency because of relative movement between the target and the detector) of the scattered return signal in order to produce an image.

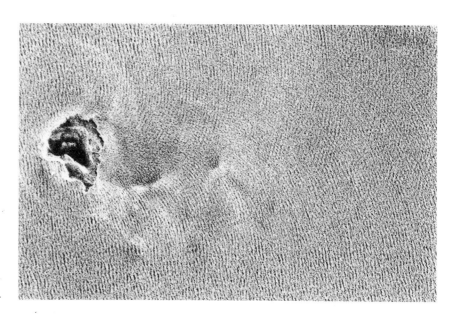

Figure 1.19 Swell-wave pattern around the island of Foula (70 km NW of Fair Isle, NE Atlantic) obtained by synthetic aperture radar (SAR). Area ≈ 30 km × 19 km.

SAR images often show a wave-like pattern (Figure 1.19), but this pattern is not a direct indication of wavelength. The regular variations in reflectivity can, however, be related to the characteristics of larger (swell) waves from which an idea of their wavelength and direction of propagation can be obtained. The precise interpretation of SAR images is hampered by such difficulties as:

(i)   SAR assumes a moving satellite and a stationary target, whereas both waves and water may be moving. This motion complicates the return signal, so that elaborate procedures have to be applied to render the image into an interpretable form.

(ii)   Image 'roughness' is affected by the entire spectrum of waves present, not just by the swell waves. Even capillary waves of a few millimetres wavelength influence the image, and hence complicate its interpretation.

(iii)   White-capping tends to scatter incident radar randomly, thus obscuring the more regular back-scatter obtained from smoother surfaces. If waves actually break, the foam causes a significant absorption of the radar signals, reducing both clarity and contrast in the final image.

3 Scatterometry

Radar scatterometers are used to analyse the strength and polarization of radar echoes, from which surface wind strength and direction can be determined. The method is complex and relies on the different back-scattering properties of wind-generated surface waves when viewed obliquely in upwind, downwind and crosswind directions.

4 Photography

Changes in the amount of reflected sunlight correlate with local roughness and wave steepness, thus revealing wave patterns on photographs. Even internal waves can be detected by photography because of their effect on surface roughness. Figure 1.20 was taken from a manned spacecraft, with a hand-held camera, and shows some internal waves in the South China Sea.

Figure 1.20(a)   Internal waves in the South China Sea (Hainan Island visible beneath clouds on lower left). Four wave packets are visible.
(b)   Tidally generated internal waves propagating into the Mediterranean from the Straits of Gibraltar. The internal waves which have amplitudes of the order of 50 feet are visible because of variations in the roughness of the sea-surface. Area = 74 km × 74 km.

In spite of the difficulties and expense, remote-sensing techniques provide a unique elevated view of the wide expanses of the oceans, and offer the only practical method of obtaining the widespread and repetitive observations that are necessary if reliable predictions of the behaviour of the sea-surface are to be made.

1.7 SUMMARY OF CHAPTER 1

1   Idealized waves of sinusoidal form have wavelength (length between successive crests), height (vertical difference between trough and crest), steepness (ratio of height to length), amplitude (half the wave height), period (length of time between successive waves passing a fixed point) and frequency (reciprocal of period). Waves transfer energy across material without significant *overall* motion of the material itself, but individual particles are displaced from, and return to, equilibrium positions as each wave passes. Surface waves occur at interfaces between fluids, either because of relative movement between them, or because the fluids are disturbed by an external force. Waves occurring at interfaces between oceanic water layers are called internal waves. Water waves, once initiated, are maintained by surface tension and gravitational force, although only the latter is significant for water waves over 1.7 cm wavelength.

2   Most sea-surface waves are wind-generated. They obtain energy from pressure differences resulting from the sheltering effect provided by the wave crests. The stronger the wind, the larger the wave, so variable winds produce a range of wave sizes. A constant wind speed produces a fully developed sea, with waves of $H_{1/3}$ (average height of highest 33% of the waves) characteristic of that wind speed. The Beaufort Scale relates sea state and $H_{1/3}$ with the causative wind speed. Water waves show cyclical variations in water level (displacement), from $-a$ (amplitude) in the trough to $+a$ at the crest. Displacement varies not only in space (one wavelength between successive crests) but also in time (one period between successive crests at one location). Steeper waves depart from the simple sinusoidal model, and more closely resemble a trochoidal wave-form.

3   Water particles in waves over deep water follow almost circular paths, but with a small net forward drift. Path diameters at the surface correspond to wave heights, but decrease exponentially with depth. In shallow water, the orbits become flattened near the sea-bed. Wave speed equals wavelength/period (or angular frequency/wave number), and is influenced both by wavelength and by depth. However, for waves in water deeper than 1/2 wavelength, wave speed is proportional to the square root of the wavelength, and is unaffected by depth. For waves in water shallower than 1/20 wavelength, wave speed is proportional to the square root of the depth, and does not depend upon the wavelength. For idealized water waves, the three characteristics c, L, and T are related by the equation $c = L/T$. In addition, each can be expressed in terms of each of the other two. For example $c = 1.56T$ and $L = 1.56T^2$.

4   Waves of different sizes become dispersed, because those with greater wavelengths and longer periods travel faster than smaller waves. If two wave trains of similar wavelength and amplitude travel over the same sea area, they interact. Where they are in phase, displacement is doubled, whereas where they are out of phase, displacement is extinguished. A single wave train results, travelling as a series of wave groups, each separated from adjacent groups by an almost wave-free region. Wave group speed in deep water is half the average speed of the two component wave trains. In shallowing water, wave speed approaches group speed, until the two coincide at depths less than 1/20 wavelength.

5   Wave energy is proportional to wave height squared, and is propagated at group speed. Total energy is conserved in a given length of wave crest, so waves entering shallowing water increase in height as their group speed falls. Wave power is wave energy propagated per second per unit length of wave crest (or wave speed multiplied by wave energy per unit area). Wave energy has been successfully harnessed on a small scale, but large scale utilization involves some environmental and navigational problems, and huge capital outlay.

6   Dissipation of wave energy (attenuation of waves) results from white-capping, friction between water molecules, air resistance, and non-linear wave–wave interaction (exchange of energy between waves of differing frequencies). Swell waves are storm-generated, travel far from their place of origin, and once established are little affected by wind or by smaller waves. Most attenuation takes place in the storm area.

7   Waves in shallow water may be refracted. Variations in depth cause variations in speed of different parts of the wave crest, which as a result

becomes refracted so as to trend parallel with bottom contours. The energy of refracted waves is conserved, so converging waves tend to increase, and diverging waves to diminish, in height. Breakers dissipate wave energy. In general, the steeper the wave and the shallower the beach, the further offshore dissipation begins. Breakers form a continuous series from steep spilling types to long-period surging breakers.

8 Waves propagating with a current have diminished heights, whereas a counter-current increases wave height, unless current speed exceeds half the wave group speed. If so, waves no longer propagate, but increase in height until they become unstable and break. Tsunamis are caused by earthquakes or by slumping of sediments, and their great wavelength means their speed is always governed by the ocean depth. Wave height is small in the open ocean, but can become destructively large near the shore. Seiches (standing waves) oscillate, so that at the antinodes there are great extremes of water level, but little lateral water movement, whereas at nodes the converse is true. The period of oscillation is proportional to container length and inversely proportional to the square root of the depth. A seiche is readily established when container length is a simple multiple of one-quarter of the seiche wavelength.

9 Waves are measured by a variety of methods, e.g. pressure gauges on the sea-floor, accelerometers in buoys on the sea-surface, and remote-sensing from satellites.

Now try the following questions to consolidate your understanding of this Chapter.

QUESTION 1.21 A wave of period 10 seconds approaching the shore has a height of 1m in deep water. Calculate:

(a) the wave speed and group speed in deep water;

(b) the wave steepness in deep water;

(c) the wave power per metre of crest in deep water;

(d) the wave power per metre of crest in water 2.5m deep.

QUESTION 1.22 A wave system consisting of short waves (wavelength 6m), together with a swell of period 22s, propagates through a narrow inlet, in which a current of 3 knots ($1.54\,\mathrm{ms^{-1}}$) runs, counter to the direction of wave propagation. Describe the wave characteristics:

(a) in the narrow inlet;

(b) at a point where the waves have passed beyond the narrow inlet into a region where the current is negligible.

(Assume the water is very deep at all the locations described.)

QUESTION 1.23 The *Ramapo* (refer back to report of waves 34m in height, Section 1.5.2) was a tanker 146m long. Assume the ship was steaming at a reduced speed of 10 knots ($5.14\,\mathrm{ms^{-1}}$), and that the wave crests took 6.3s to pass the ship from stern to bow.

(a) What was the wave speed?

(b) What was the wave steepness?

(c) How does the wave period, consistent with the answers to (a) and (b), compare with the wave period of 14.8s reported by the *Ramapo*?

(d) What would be the implications for maximum wave steepness of a wave period of 14.8s?

(e) Comment upon any inconsistencies revealed.

QUESTION 1.24   What sort of waves would you expect to see on a beach of intermediate slope:

(a) after a prolonged spell of calm weather?

(b) during a severe gale, with a Force 9 wind blowing onshore?

CHAPTER 2	TIDES

'. . . being governed by the watery Moon . . .'
Richard III, Act II, Sc. II.

The longest oceanic waves are those associated with the tides, and are characterized by the rhythmic rise and fall of sea-level over a period of several hours. The rising tide is usually referred to as the **flow**, whereas the falling tide is called the **ebb**. This ebb and flow of the tide, which can be very powerful, causes tidal currents (see Section 2.4.1), especially where the water is constrained by shallow depth or adjacent land masses. From the earliest times it has been realized that there is some connection between the tides and the Sun and Moon. Tides are highest when the Moon is full or new, and the times of high tide at any given location can be approximately (but not exactly) related to the position of the Moon in the sky. Because the relative motions of the Earth, Sun and Moon are complicated, it follows that their influence on tidal events results in an equally complex pattern. Nevertheless, the magnitudes of the tide-generating forces can be precisely formulated, although the response of the oceans to these forces is modified by the permanent effects of topography and the transient effect of weather patterns.

2.1 TIDE-PRODUCING FORCES—THE EARTH–MOON SYSTEM

The Earth and the Moon form a single system, mutually revolving around a common centre of mass, with a period of 27.3 days. The orbits are in fact slightly elliptical, but to simplify matters we will treat them as circular for the time being. The Earth revolves eccentrically about the common centre of mass, which means that all points within and upon the Earth follow circular paths, all of which have the same radius (two examples are shown on Figure 2.1 as points C and X). Each point will also have the same angular velocity of $2\pi/27.3$ days. Because the angular velocities, and the radii of the circular paths travelled, are the same for all points, it follows that all points on and within the Earth experience an equal acceleration (the product of the radius and the square of the angular velocity) and hence an equal centrifugal force as a result of this eccentric motion.

The eccentric motion has nothing whatsoever to do with the Earth's rotation (spin) upon its own axis, and should not be confused with it, nor should the centrifugal force due to the eccentric motion be confused with the centrifugal forces due to the Earth's spin which increase with distance from the axis, whereas those due to the eccentric motion are equal at all points on Earth.

If you find this concept difficult, the following simple analogy may help. Imagine you are whirling a small bunch of keys on a short length (say 25 cm) of chain. The keys represent the Moon, and your hand represents the Earth. You are rotating your hand eccentrically (but unlike the Earth it is not spinning as well), and all points on and within your hand are

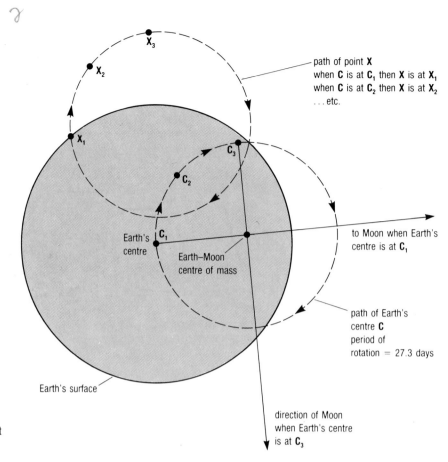

Figure 2.1 The eccentric revolution of the Earth about the Earth–Moon centre of mass, as viewed from above one of the poles, when the Moon is directly above the Equator. Every point on Earth follows a circular path analogous to those traced out by points C and X.

experiencing the same angular velocity and the same centrifugal force. Provided your bunch of keys is not too large, the centre of mass of the 'hand-and-key' system lies within your hand.

The total centrifugal force within the Earth–Moon system exactly balances the forces of gravitational attraction between the two bodies, so that the system as a whole is in equilibrium, i.e. we should neither lose the Moon, nor collide with it, in the near future. The centrifugal forces are directed parallel to a line joining the centres of the Earth and the Moon (see Figure 2.2). Now consider the magnitude of the gravitational force exerted by the Moon on the Earth. This will not be the same at all points on the Earth's surface, because not all these points are the same distance away from the Moon. So, points on the Earth nearest the Moon will experience a greater gravitational pull from the Moon than will points on the opposite side of the Earth. Moreover, the direction of the Moon's gravitational pull at all points will be directed towards the centre of the Moon, and hence, except on the line joining the centres of the Earth and Moon, will not be exactly parallel to the direction of the centrifugal forces. The resultant (i.e. the composite effect) of the two forces is known as the **tide-producing force**, and, depending upon its position on the Earth's surface with respect to the Moon, may be directed into, parallel to, or away from, the Earth's surface. The relative strengths and directions (not strictly to scale) of the forces involved are shown on Figure 2.2.

3

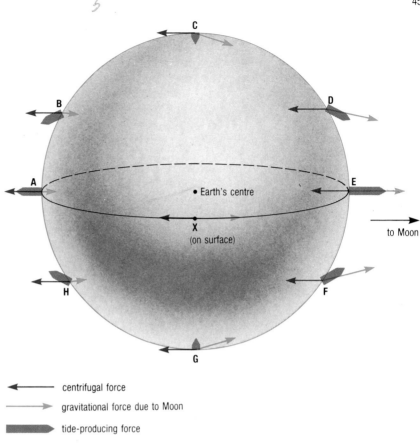

C

B

D

A

• Earth's centre

E

X
(on surface)

to Moon

H

F

G

⟵ centrifugal force

⟶ gravitational force due to Moon

⟹ tide-producing force

Figure 2.2 The derivation of the tide-producing forces (not to scale). The centrifugal force has exactly the same magnitude and direction at all points, whereas the gravitational force exerted by the Moon on the Earth varies in both magnitude (inversely with the square of the distance from the Moon) and direction (directed towards the Moon's centre, with the angles exaggerated for clarity). The tide-producing force at any point is the *resultant* of the gravitational and centrifugal forces at that point, and varies inversely with the cube of the distance from the Moon (see text).

QUESTION 2.1   What would be the direction and approximate magnitude (within the context of Figure 2.2) of the tide-producing forces at:

(a)   a point on the Earth's surface represented by point X on Figure 2.2?

(b)   the Earth's centre?

The gravitational force (F_g) between two bodies is given by:

$$F_g = \frac{GM_1M_2}{R^2} \tag{2.1}$$

where M_1 and M_2 are the masses of the two bodies, R is the distance between their centres, and G is the universal gravitational constant ($6.672 \times 10^{11}\,\mathrm{N\,m^2\,kg^{-2}}$).

You may be puzzled at how to reconcile equation 2.1 with the statement on Figure 2.2 that the magnitude of the tide-producing force exerted by the Moon on the Earth varies inversely with the cube of the distance. Consider the point marked E on Figure 2.3. The gravitational attraction of the Moon at E (F_{gE}) is greater there than that at the Earth's centre, because E is nearer to the Moon by the distance of the Earth's radius (a).

Because the gravitational force exerted by the Moon on a point at the Earth's centre is exactly equal and opposite to the centrifugal force there, the tide-producing force at the centre of the Earth is zero. Now as the centrifugal force is equal at all points on Earth, and at the Earth's centre is equal to the gravitational force exerted there by the Moon, it follows that we can substitute the expression on the right-hand side of equation 2.1 (i.e. GM_1M_2/R^2) for the centrifugal force.

The tide-producing force at E (TPF_E) is given by the force due to gravitational attraction of the Moon at E (F_{gE}) minus the centrifugal force at E, i.e.

$$TPF_E = \frac{GM_1M_2}{(R - a)^2} - \frac{GM_1M_2}{R^2}$$

which simplifies to:

$$TPF_E = \frac{GM_1M_2a(2R - a)}{R^2(R - a)^2}$$

Now a is very small compared to R, so $2R - a$ can be approximated to $2R$, and $(R - a)^2$ to R^2, giving the approximation:

$$TPF_E \approx \frac{GM_1M_22a}{R^3} \tag{2.2}$$

The equation is slightly more complex for points on the Earth that do not lie directly on a line joining the centres of the Earth and Moon. For example, at point P on Figure 2.3(a), the gravitational attraction (F_{gP}) would be, to a first approximation:

$$F_{gP} = \frac{GM_1M_2}{(R - a \cos \psi)^2} \tag{2.3}$$

The length $a \cos \psi$ is marked on Figure 2.3(a). (ψ is the Greek 'psi'.)

Before reading on, have another look at Figure 2.2, and consider at which of the lettered points on that Figure the local tide-producing force would have most effect on the tides.

You may have considered point E as your answer. Certainly, E is nearest to the Moon, and hence is one of the two points where the difference between the centrifugal force and the gravitational force exerted by the Moon is greatest. However, all the resultant tide-producing force is acting vertically against the pull of the Earth's own gravity, which happens to be about 9×10^6 greater than the tide-producing force. Hence the local effect of the tide-producing forces at point E is negligible. Similar arguments apply at point A, except that F_{gA} is *less* than the centrifugal force, and consequently the tide-producing force at A is equal in magnitude to that at E, but directed away from the Moon (see also Figure 2.3(b)).

At the lettered points B, D, F and H on Figure 2.2 (which lie on the small circles defined on Figure 2.3(a)), the effects of the tide-producing forces are greatest, because at each there is a large horizontal component (known as the **tractive force**) of the tide-producing force. It is the tractive forces which cause the water to move, because, although small compared with the Earth's gravitational field, this horizontal component is unopposed by any other lateral force (apart from friction at the sea-bed, which is negligible in this context). Figure 2.3(b) shows where on the

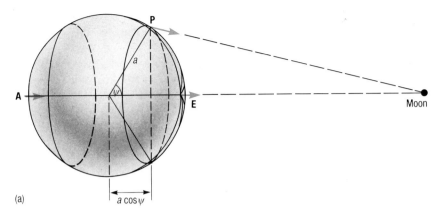

(a)

$a \cos \psi$

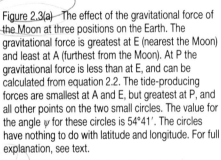

North Pole

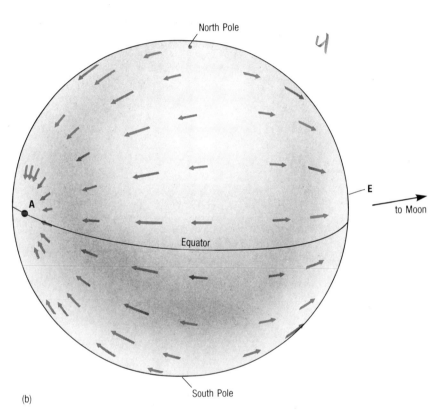

Equator

E

to Moon

South Pole

(b)

Figure 2.3(a) The effect of the gravitational force of the Moon at three positions on the Earth. The gravitational force is greatest at E (nearest the Moon) and least at A (furthest from the Moon). At P the gravitational force is less than at E, and can be calculated from equation 2.2. The tide-producing forces are smallest at A and E, but greatest at P, and all other points on the two small circles. The value for the angle ψ for these circles is 54°41′. The circles have nothing to do with latitude and longitude. For full explanation, see text.

(b) The relative magnitudes of the tractive forces at various points on the Earth's surface. The assumption is made that the Moon is directly over the Equator, i.e. at zero declination. Points A and E correspond to those on Figure 2.2.

Earth the tractive forces are at a maximum when the Moon is over the Equator. In this simplified case, the tractive forces would result in total movement of water towards points A and E on Figure 2.3(b). In other words, an equilibrium state would be reached (called the **equilibrium tide**), producing an ellipsoid with its two bulges directed towards and away from the Moon. So, paradoxically, although the tide-producing forces are minimal at A and E, those are the points towards which the water would go.

In practice, this ellipsoid does not develop, because the Earth rotates upon its own axis. The two bulges, in order to maintain the same position relative to the Moon, would have to travel around the world at the same rate (but in the opposite direction) as the Earth rotates with respect to the Moon.

Because the Moon revolves about the Earth–Moon centre of mass once every 27.3 days, in the same direction as the Earth rotates upon its own axis (which is once every 24 hours), the period of the Earth's rotation with respect to the Moon is 24 hours and 50 minutes (a **lunar day**). This is the reason why the times of high tides at many locations are almost an hour later each successive day (Figure 2.4).

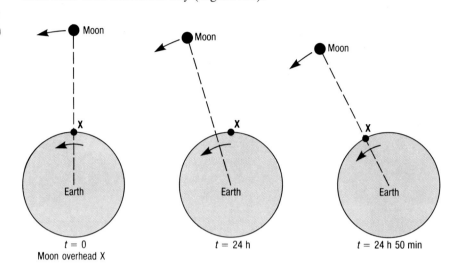

Figure 2.4   The relationship between a solar day of 24 hours and a lunar day of 24 hours and 50 minutes as seen from above the Earth's North Pole. Point X on the Earth's surface when the Moon is directly overhead comes back to its starting position 24 hours later. Meanwhile, the Moon has moved on in its orbit, so that point X has to rotate further (another 50 minutes' worth) before it is once more directly beneath the Moon.

QUESTION 2.2(a)   Using a value of 40000 km for the Earth's circumference, calculate the required speed of the tidal bulges in order to maintain an equilibrium tide around the Equator. (Assume for simplicity that the Moon is directly overhead at the Equator.)

(b)   How deep would the oceans have to be to allow the tidal bulges to travel as shallow water waves at the speed you calculated in part (a)?

The concept of the equilibrium tide was developed by Newton in the seventeenth century. Your answer to Question 2.2 shows that, in practice, an equilibrium tide cannot occur at low latitudes on Earth, and, as will be seen in Section 2.3, the actual tides behave differently. However, the theoretical equilibrium tide demonstrates the fundamental periodicity of the tides on a semi-diurnal basis of 12 hours and 25 minutes.

2.1.1   VARIATIONS IN THE LUNAR-INDUCED TIDES

The relative positions and orientations of the Earth and Moon are not constant, but vary according to a number of interacting cycles. As far as a simple understanding of the tide-generating mechanism is concerned, only two cycles have a significant effect on the tides.

1   The Moon's declination

The Moon's orbit is not in the plane of the Earth's Equator, but is inclined at an angle of 28°. This means that a line joining the centre of the Earth to that of the Moon ranges up to about 28° either side of the equatorial plane, over a cycle of 27.2 days (not to be confused with the 27.3-day period of the Earth–Moon system's rotation given in Section 2.1). To an observer on Earth, successive paths of the Moon across the sky appear to rise and fall over a 27.2 day cycle, in a similar way to the variation of the Sun's apparent path over a yearly cycle. When the Moon is at a large angle of declination, the plane of the two tidal bulges will be offset with respect to the Equator, and their effect at a given latitude will be unequal, particularly at mid-latitudes. Hence the heights reached by the semi-diurnal (i.e. twice daily) tides will show diurnal (i.e. daily) inequalities (Figure 2.5).

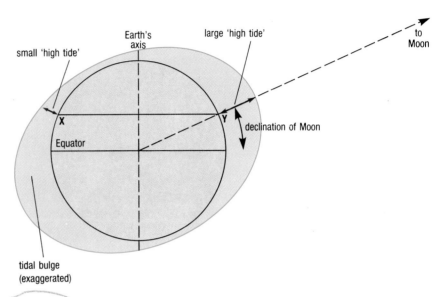

Figure 2.5   The production of unequal tides (tropic tides—see text) at mid-latitudes due to the Moon's declination. An observer at Y would experience a higher high tide than would an observer at X. 12 hours and 25 minutes later their positions would be reversed, i.e. each observer would notice a diurnal inequality.

QUESTION 2.3   What will be the extent of the diurnal tidal inequality due to the Moon as seen by an observer on the Tropic of Capricorn (and within sight of the sea) roughly seven days after the situation shown in Figure 2.5?

Your answer to Question 2.3 emphasizes the cyclical nature of this variation. At maximum declination of 28° the Moon is (roughly) over one of the 'Tropics' (latitude 23°), the diurnal variation is greatest, and the tides are known as **tropic tides**; whereas at minimum (zero) declination (when the Moon is vertically above the Equator), there is no diurnal variation and the tides are called **equatorial tides**.

2   The Moon's elliptical orbit

The orbit of the Moon around the Earth–Moon centre of mass is not circular but elliptical. The consequent variation in distance from Earth to Moon results in corresponding variations in the tide-producing forces. When the Moon is closest to Earth, it is said to be in **perigee**, and the Moon's tide-producing force is increased by up to 20% above the average value. When the Moon is furthest from Earth, it is said to be in **apogee**, and the tide-producing force is reduced to about 20% below the average value. The interval between successive perigees is 27.5 days.

2.2 TIDE-PRODUCING FORCES—THE EARTH–SUN SYSTEM

The Sun also plays its part as a tide-raising agent. Just as the Moon does, the Sun produces tractive forces and two equilibrium tidal bulges. The magnitude of the Sun's tide-producing force is about 0.46 that of the Moon, because, although enormously greater in mass than the Moon, the Sun is some 360 times further from the Earth. As we saw in Section 2.1, tide-producing forces vary directly with the mass of the attracting body, but are inversely proportional to the cube of its distance from Earth. The two solar equilibrium tides produced by the Sun sweep westwards around the globe as the Earth spins towards the east. The solar tide thus has a semi-diurnal period of twelve hours.

Just as the relative heights of the two semi-diurnal lunar tides are influenced by the Moon's declination, so there are diurnal inequalities in the solar-induced components of the tides because of the Sun's declination. The Sun's declination varies over a yearly cycle, and ranges 23° either side of the equatorial plane.

QUESTION 2.4   At what time(s) of the year will the solar-induced component of the tide show maximum diurnal inequality?

As in the case of the Moon's orbit round the Earth, the orbit of the Earth around the Sun is elliptical, with a consequent minimum Earth–Sun distance, when the Earth is said to be at **perihelion**, and a maximum distance, when it is said to be at **aphelion**. However, the difference in distance between perihelion and aphelion is only about 4%, compared with an approximate 13% difference between perigee and apogee.

2.2.1   INTERACTION OF SOLAR AND LUNAR TIDES

In order to understand the interaction between solar and lunar tides, it is helpful to consider the simplest case, where the declinations of the Sun and Moon are both zero. Figure 2.6 shows these conditions, looking down on the Earth from above the North Pole. The direction of rotation of the Earth is shown arrowed, and the solar and lunar tides are shown diagrammatically. The complete cycle of events takes 29.5 days.

In Figure 2.6(a) and 2.6(c), the tide-generating forces of the Sun and Moon are acting in the same directions, and the solar and lunar equilibrium tides coincide. The tidal range produced is large, i.e. the high tide is higher, and the low tide is lower, than the average. Such tides are known as **spring tides**. When spring tides occur, the Sun and Moon are said to be either in **conjunction** (at new Moon—Figure 2.6(a)) or in **opposition** (at full Moon—Figure 2.6(c)). There is a collective term for both situations: the Moon is said to be in **syzygy** (pronounced 'sizzijee').

In Figure 2.6(b) and (d), the Sun and Moon act at right angles to each other, and the solar and lunar tides are out of phase. The tidal range is correspondingly smaller than average. These tides are known as **neap tides**, and the Moon is said to be in **quadrature** when neap tides occur. Inshore fishermen often refer to spring and neap tides by the descriptive names of 'long' and 'short' tides respectively.

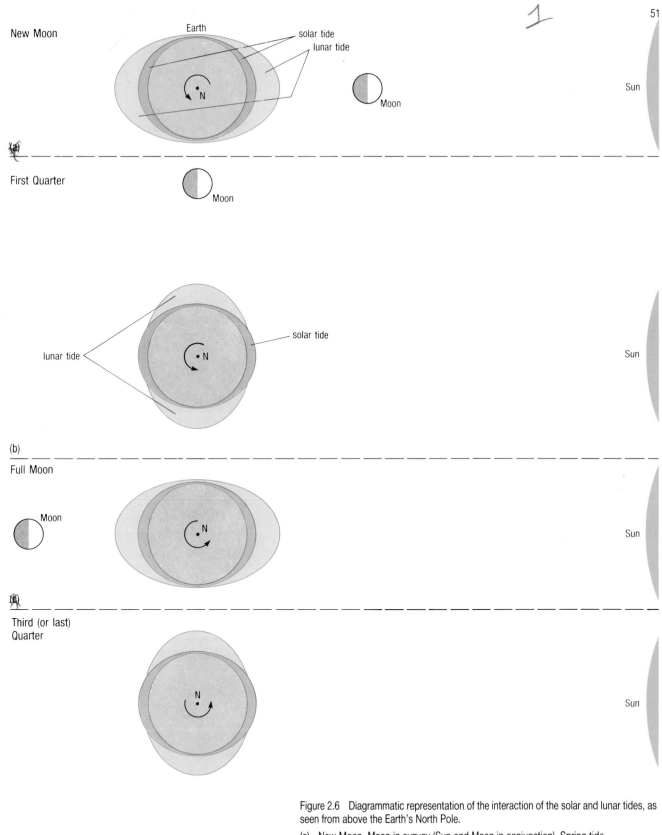

Figure 2.6   Diagrammatic representation of the interaction of the solar and lunar tides, as seen from above the Earth's North Pole.

(a)   New Moon. Moon in syzygy (Sun and Moon in conjunction). Spring tide.

(b)   First quarter. Moon in quadrature. Neap tide.

(c)   Full Moon. Moon in syzygy (Sun and Moon in opposition). Spring tide.

(d)   Third (or last) quarter. Moon in quadrature. Neap tide.

QUESTION 2.5(a)   What is the time interval between two successive neap tides?

(b)   What is the tidal state 22 days after the Moon is in syzygy?

(c)   How soon after the new Moon might a tide of 'average' range be expected?

(d)   If the simplification adopted in Figure 2.6 (of zero declination for both Sun and Moon) were literally true, what astronomical phenomena would be observed on the Earth's Equator if the Sun, Moon and Earth were in the positions shown in Figure 2.6(a) and (c) respectively?

The regular changes in the declinations of the Sun and Moon, and their cyclical variations in position with respect to the Earth, produce very many harmonic constituents, each of which contributes to the tide at any particular time and place. One interesting situation is the 'highest astronomical tide', i.e. that which would create the greatest possible tide-raising force, with the Earth at perihelion, the Moon in perigee, the Sun and Moon in conjunction and both Sun and Moon at zero declination. Such a rare combination would produce a tidal range greater than normal. For example, the normal tidal range at Newlyn, Cornwall, is about 3.5m, the mean spring tidal range about 5m, and the highest astronomical tidal range about 6m. However, there is no immediate need to sell any seaside property which you may own—the next such event is not due until about AD 6580.

2.3   THE DYNAMIC THEORY OF TIDES

Newton, in formulating the equilibrium theory of tides, was well aware of discrepancies between the predicted equilibrium tides and the observed tides, but did not pursue the matter any further. There are a number of reasons why actual tides do not behave as equilibrium tides:

1   The average depth of the oceans is much less than the 20km you calculated as the answer to Question 2.2(b). Assuming a depth over the abyssal plains of about 5500m, the speed of any wave longer than a few kilometres is limited to about $230\,\mathrm{m\,s^{-1}}$ in the open ocean, and is less in shallower seas (equation 1.5).

2   The presence of land masses prevents the tidal bulges from directly circumnavigating the globe, and the shape of the ocean basins constrains the direction of tidal flows.

3   The rate of rotation of the Earth on its axis is too rapid for the inertia of the water masses to be overcome in sufficient time to establish an immediate equilibrium tide. A time-lag in the oceans' response to the tractive forces is inevitable—and this is fortunate because otherwise each high tide would arrive in the same way as an outsized tsunami.

4   Lateral water movements induced by tide-generating forces are subject to the **Coriolis force**, which deflects tidal flows *cum sole* (*cum sole* literally means *with the Sun*, i.e. to the right, or clockwise, in the Northern Hemisphere, and to the left, or anticlockwise, in the Southern Hemisphere).

The **dynamic theory of tides** was developed during the eighteenth century by scientists and mathematicians such as Bernoulli, Euler and Laplace. They attempted to understand tides by considering ways in which the depths and configurations of the ocean basins, the Coriolis force, inertia, and frictional forces might influence the behaviour of fluids subjected to rhythmic forces. As a consequence, the dynamic theory of tides is intricate, and solutions of the equations are complex. Nevertheless, the dynamic theory has been steadily refined, and theoretical tides can now be computed which are very close approximations to the observed tides. The dynamic theory of tides is best understood by considering the simplest situation, where Sun and Moon are both at zero declination, and in syzygy, so that solar and lunar tides coincide. We then have only one equilibrium tide to think about. The answer to Question 2.2 demonstrated that an ocean depth greater than 20 km would be required for the theoretical equilibrium tide at the Equator to 'keep up with' the Moon's passage around the Earth. Because the oceans are everywhere less than 20 km deep, the actual tide will be retarded with respect to the equilibrium tide, i.e. it will 'lag'.

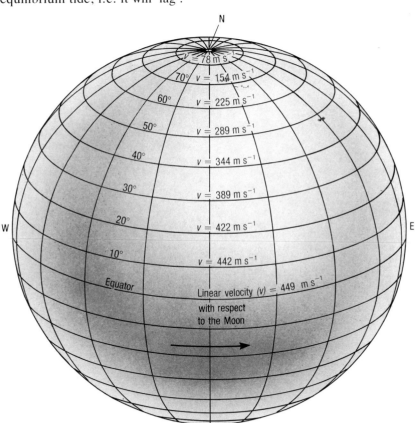

Figure 2.7 Linear velocities (v) at various latitudes of the Earth's surface with respect to the Moon.

Figure 2.7 shows, at various latitudes, linear velocities (v) of the Earth's surface with respect to the Moon. Compare these velocities with the speed at which a theoretical equilibrium tide could propagate as a shallow-water wave over an abyssal plain of 5.5 km depth (roughly at 230 m s^{-1}). However, the oceans are not 5.5 km deep everywhere—the average depth is somewhat less, so the theoretical speed of the tidal wave will be less.

QUESTION 2.6 For this question, assume an average ocean depth of 4080 m (to simplify the calculations).

(a) What will be the tidal lag, in hours, at a latitude of 26°?

(b) What will be the tidal lag, in hours, at a latitude of 10°?

54

From the answer to Question 2.6 it can be seen that at low latitudes the tides will lag 90° of longitude behind the theoretical equilibrium tide. This phenomenon extends either side of the Equator to those latitudes where the Earth's surface (with respect to the Moon) reaches a linear velocity equal to twice the speed at which tides could propagate across the oceans. At these lower latitudes, the lag is limited to 6 hours 12 minutes, so that high tides occur 6 hours 12 minutes and 18 hours 36 minutes after the Moon's passage overhead. Such tides are called **indirect tides**.

At latitudes above about 26°, the tidal lag is less than 6 hours 12 minutes. The precise lag is always constant for any one location, but decreases with increasing latitude, until there is zero lag at about latitude 65°. Consider the speed of the equilibrium tide around the Antarctic Circle (66.5°S). The distance around the circle is some 17300 km, and the linear velocity is thus about 190ms^{-1}, which is less than the velocity of the tidal wave you assumed when answering Question 2.6(a). Therefore there is no lag, and the actual tide would have no difficulty in keeping up with the theoretical lunar equilibrium tide, and, in theory, high tides would occur at (and 12

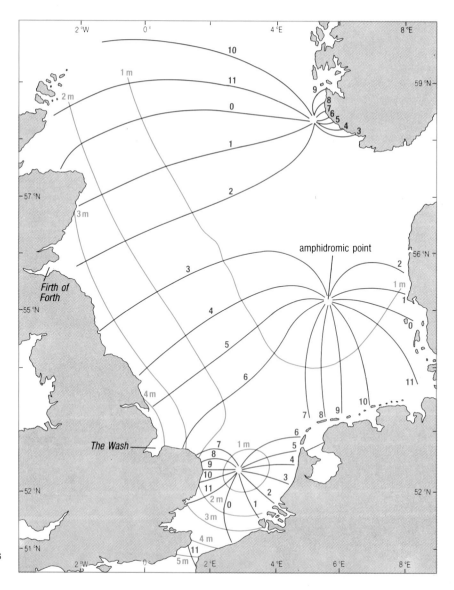

Figure 2.8 Amphidromic systems in the North Sea. The figures on the co-tidal lines indicate the time of high water in 'lunar hours' (i.e. 1/24 of a lunar day of 24.8 hours = about 1 hour and 2 minutes) after the Moon has passed the Greenwich meridian. Blue lines are co-range lines and red lines are co-tidal lines.

hours 25 minutes after) the Moon's passage. Tides of this nature are called **direct tides**. According to the dynamic theory, then, all tides at low latitudes (less than about 26°) would be indirect tides, and all tides at high latitudes (more than about 65°) would be direct. However, tides do not necessarily ebb and flow parallel with the Equator, as the dynamic theory assumes. A considerable longitudinal component of flow occurs, and the actual tidal pattern is thus more complicated than the simple dynamic theory.

So far, we have ignored the configuration of the ocean basins. In fact, the combined constraint of ocean basin geometry and the influence of the Coriolis force results in the development of **amphidromic systems**, in each of which the crest of the tidal wave at high water circulates around an **amphidromic point** once during each tidal period (Figures 2.8 and 2.9). The tidal range is zero at each amphidromic point, and increases outwards away from it.

In each amphidromic system, **co-tidal lines** can be defined, which link all the points where the tide is at the same stage (or phase) of its cycle. The co-tidal lines thus radiate outwards from the amphidromic point.

Cutting across co-tidal lines, approximately at right angles to them, are **co-range lines**, which join places having an equal tidal range. Co-range lines form more or less concentric circles about the amphidromic point, representing larger and larger tidal ranges the further away they are from it. Figure 2.8 shows the amphidromic systems for the North Sea, and Figure 2.9 shows the computed world-wide amphidromic systems for the dominant tidal component (see Section 2.3.1).

Figure 2.9 Computer-generated diagram of world-wide amphidromic systems for the dominant semi-diurnal lunar tidal component (see Table 2.1). Co-tidal lines are in red and co-range lines are in blue.

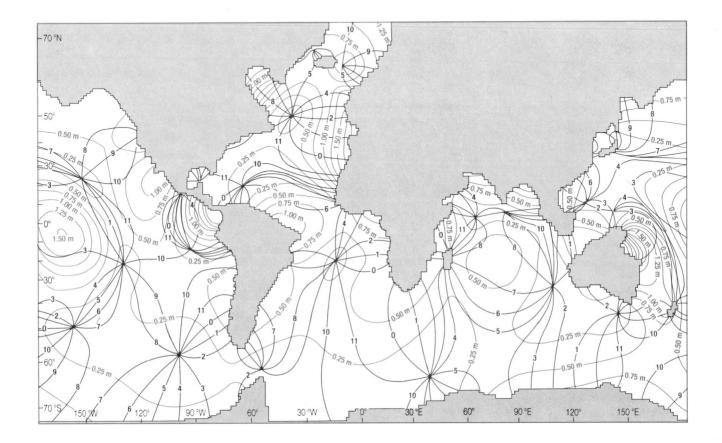

QUESTION 2.7 Assume that a high tide coincides with the co-tidal lines marked zero (i.e. '00') on Figure 2.8. At what stage of the tidal cycle is:

(a)   The Wash?

(b)   The Firth of Forth?

(c)   Which one of (a) and (b) has the greater tidal range?

Inspection of Figures 2.8 and 2.9 shows that, with a few exceptions, the tidal waves of amphidromic systems tend to rotate anticlockwise in the Northern Hemisphere and clockwise in the Southern Hemisphere. At first sight, the pattern of rotation appears to conflict with the principle that the Coriolis force deflects moving fluid masses *cum sole*. However, consider the bay shown in Figure 2.10.

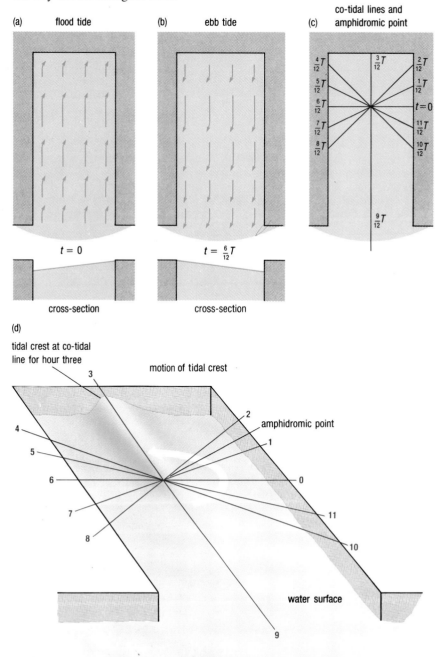

Figure 2.10   The development of an amphidromic system. The hypothetical bay shown is in the Northern Hemisphere.

(a)   Flood tide. Water is deflected to the right by the Coriolis force, i.e. towards the east bank.

(b)   Ebb tide. Returning water is deflected to the right by the Coriolis force, i.e. towards the west bank.

(c)   An amphidromic system is established.

(d)   The tidal wave travels anticlockwise.

In Figure 2.10(a) the flooding tide is deflected to the right by the Coriolis force (the bay is in the Northern Hemisphere), and the water is piled up on the eastern side of the bay. Conversely, when the tide ebbs, the water becomes piled up on the western side (Figure 2.10(b)). Hence, because the tidal wave is constrained by land masses, an anticlockwise amphidromic system is set up (Figure 2.10(c) and (d)). It is noteworthy that the main exceptions to the general pattern of rotation shown on Figure 2.9 are systems which are not constrained by land masses (e.g. the South Atlantic, Mid-Pacific and North Pacific amphidromic systems) or where the system rotates about an island (e.g. Madagascar, Ceylon and New Zealand).

Confined amphidromic systems are a type of **Kelvin wave**, in which the restoring force (gravity) is reinforced near the coasts by forces generated by the head of water associated with the sloping sea-surface (Figure 2.10(a) and (b)). Kelvin waves occur where the deflection caused by the Coriolis force is either constrained (as at coasts) or is zero (as at the Equator). Because the Coriolis force acts *cum sole*, Kelvin waves can only travel eastwards at the Equator.

2.3.1 PREDICTION OF TIDES BY THE HARMONIC METHOD

The harmonic method is the most usual and satisfactory method for the prediction of tidal heights. It makes use of the knowledge that the observed tide is the sum of a number of components or **partial tides**, each of whose periods precisely corresponds with the period of one of the relative astronomical motions between Earth, Sun and Moon. Each of the partial tides has an amplitude and phase which is unique to a given location. In this context, phase means the fraction of tidal cycle that has been completed at a given reference time. It depends upon the period of the tide-raising force concerned, and upon the lag of the partial tide for that particular location.

The determination of amplitude and phase for each partial tide at a particular point, such as a seaport, requires a record of tidal heights obtained over a time that is long compared with the periods of the partial tides concerned. As many as 390 components have been identified. Table 2.1 shows four semi-diurnal, three diurnal and two longer-period components.

Table 2.1 Some principal tidal components.

Name of tidal component	Symbol	Period in solar hours	Coefficient ratio ($M_2 = 100$)
Principal lunar	M_2	12.42	100
Principal solar	S_2	12.00	46.6
Larger lunar elliptic	N_2	12.66	19.2
Luni–solar semi-diurnal	K_2	11.97	12.7
Luni–solar diurnal	K_1	23.93	58.4
Principal lunar diurnal	O_1	25.82	41.5
Principal solar diurnal	P_1	24.07	19.4
Lunar fortnightly	M_f	327.86	17.2
Lunar monthly	M_m	661.30	9.1

Even using these few major components, the production of a tide-table for a port for an entire year used to be a very time-consuming activity. In the early years of harmonic analysis, they were computed by hand. The first machine to do the job was invented by Lord Kelvin in 1872. Electronic computers are admirably suited to this repetitive procedure, and tide-tables produced for ports all over the world now take little time to prepare.

The precision achieved by radar altimeters (Section 1.6.1) is such that tidal ranges in the deep oceans can be measured. Information on tidal amplitude and phase was extracted from the *Seasat* data (collected in 1978) by oceanographers working at the Institute of Oceanographic Sciences, Bidston, and found to be in good agreement with predicted values.

2.4 TYPES OF TIDE

Having examined the theory, let us see how actual tides behave in different places and how the different types are classified. The simplest way of classifying tides is by the dominant period of the observed tide. This is based on the ratio (F) of the sum of the amplitudes of the two main diurnal components (K_1 and O_1) to the sum of the amplitudes of the two main semi-diurnal components (M_2 and S_2).

QUESTION 2.8(a)   From Figure 2.11, what are the main differences between tidal cycles characterized by high and low values of the ratio F?

(b)   Would you expect the interval between spring tides to be 14.75 days at all times, and at all locations, irrespective of the other types of tidal fluctuation?

A high value of F (say above 3.0) implies a diurnal tidal cycle, i.e. only one high tide occurs daily, and fluctuations in the tidal range are largely due to changes in the Moon's declination. Tides are very small at times of zero lunar declination. Low values of F (say less than 0.25) imply a semi-diurnal tide, and the main fluctuations in tidal range are due to the relative positions of Sun and Moon, giving a spring–neap variation.

Between the two extremes are the mixed tidal types, where the daily inequality caused by the declination of the Moon (see Section 2.1.1(cycle 1)) is important, and variations in the amplitudes of, and time intervals between, successive high tides can be considerable. The middle two tidal records on Figure 2.11 show diurnal inequalities where typical 'large tides' alternate with 'half-tides'. The time intervals between successive high waters are unequal when the half-tide is intermittent, as at Manila. Note the change from tropic tides at days 0–6 to equatorial tides at days 7–12 in the cases of both Manila and San Francisco (see Section 2.1.1).

Local effects can modify these basic patterns, particularly the local effects of harmonics (i.e. simple multiples of the frequency) of the partial tides. For example, the quarter-diurnal component M_4 (twice the frequency of M_2, the semi-diurnal or principal lunar component) and the one-sixth-diurnal component M_6 (three times the frequency of M_2) are generated in addition to the semi-diurnal component. In most locations, the effect of the two harmonics is insignificant compared with the principal

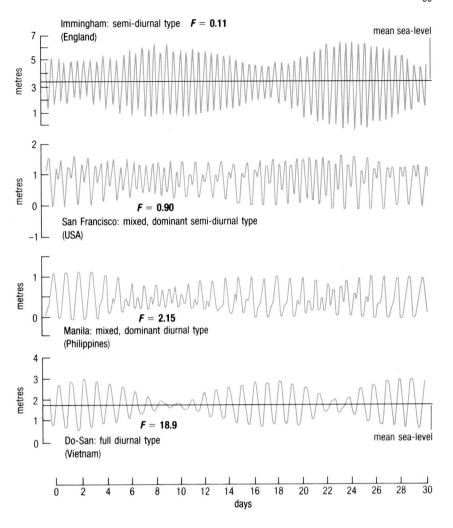

Figure 2.11 Examples of different types of tide curve, in England, the USA, the Philippines, and Vietnam. For full explanation, see text.

component, but along the Dorset and Hampshire coasts of the English Channel each has a larger amplitude than usual. Moreover, the two harmonics are in phase, and their combined amplitude is significant when compared to that of M_2. (Just west of the Isle of Wight, M_2 is about 0.5m, M_4 about 0.15m, and M_6 about 0.2m). The additive effect of all three components largely contributes to the double high waters at Southampton and the double low waters at Portland. There is no truth in the popular myth that double high water at Southampton is caused by the tide flooding at different times around either end of the Isle of Wight.

2.4.1 TIDES AND TIDAL CURRENTS IN SHALLOW SEAS

The vertical water movements associated with the rise and fall of the tide are accompanied by horizontal water motions termed tidal currents. These tidal currents have the same periodicities as the vertical oscillations, but tend to follow an elliptical path and do not normally involve a simple to-and-fro motion (Figure 2.12(a)).

(a)

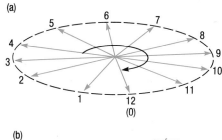

(b)

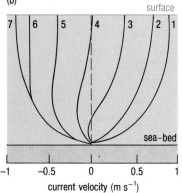

Figure 2.12(a)   The elliptical path followed by water particles in a tidal current during a complete tidal cycle. The successive directions of the current are shown by arrows. The length of the arrows is proportional to current velocity at the relevant time. Numbers refer to lunar hours (62 minutes) measured after an arbitrary starting time in the cycle.

(b)   A series of vertical tidal current profiles, showing retardation of the currents close to the sea-bed. The numbers refer to time in lunar hours after an arbitrary starting time, and only half a tidal cycle is shown.

The sense of rotation of the ellipse may be either clockwise or anticlockwise, but rotations *cum sole* tend to be favoured if there are no constraining land masses. In small embayments, the effect of the Earth's rotation is insignificant, but the frictional effect of the sea-bed and the constraining effect of the land masses upon the current cannot be neglected. In large basins such as the North Sea, the rotational effect is more important than the frictional effect. Figure 2.12(b) shows a series of current profiles during a tidal cycle. The retardation of the flow near the bottom of the profile is typical of tidal current behaviour in shallow seas. The form of such profiles can be important when considering the movement and distribution of sediments.

In areas where the tidal current is strong enough, the frictional drag at the sea-bed produces a vertical current shear, and the resultant turbulence causes vertical mixing of the lower water layers. In other areas, where tidal currents are weaker, little mixing occurs, and thus stratification (water layers of different densities) can develop. The boundaries (fronts) between such contrasting areas of mixed and of stratified waters are often steeply inclined and sharply defined, so that there are marked lateral differences in water density either side of the front.

If the tide behaves as an ordinary progressive wave, then maximum currents occur at high and low tides. On the other hand, if reflection of the progressive wave results in a standing wave (see Section 1.5.4) being established in a basin, then one end of the basin will experience high water while the other end is at low water. The flow will be directed from the end at which the level is falling to the end at which it is rising, and the maximum flow rate will be at 'mid-tide', when both ends are at the same level. In practice, however, the maximum currents will coincide neither with high/low tides, nor with mid-tides, but will be somewhere in between. For example, in the case of the North Sea, the tidal oscillations are partly determined by the dimensions of the North Sea basin (which has a natural period of about 40 hours) and partly by the progressive semi-diurnal tides entering from the Atlantic. As a result, a standing wave with three nodes tends to develop in the North Sea. The water is deflected by the Coriolis force, and forms three amphidromic systems (as shown in Figure 2.8) by the same mechanism that was outlined in Section 2.3 and Figure 2.10.

In the case of the Bay of Fundy, Nova Scotia, the natural period of oscillation is about 12.5 hours, i.e. it is close to that of the semi-diurnal tide. As a result, there is a strong resonant oscillation, a tidal range of some 15 m at the head of the bay, and a strong tidal current at mid-tide.

QUESTION 2.9   The Bay of Fundy is about 270 km long, and has an average depth of about 60 m. Are these dimensions consistent with the resonant period given in the last paragraph?

Strong tidal currents can also be produced by local constrictions, such as in narrow straits between two seas (e.g. the Straits of Dover). Such currents are known as hydraulic currents and result from the hydraulic pressure gradients caused by differences in sea-level at either end of the straits. The relatively steep gradient of the water surface may result in a tidal race of several knots. Where the French coast reaches out towards the Channel Island of Alderney, spring tides can routinely generate currents of 10 knots (5.14 m s^{-1}). Interacting currents can result in confused seas—even white-capping—on an otherwise calm day. The sea off the tip of Portland Bill, Dorset, can present similar problems. Such areas are marked on navigational charts as overfalls, and are particularly to be avoided when waves are steepened by opposition to such tidal currents.

2.4.2 STORM SURGES

An additional complication in the prediction of tidal heights is that meteorological conditions can considerably change the height of a particular tide, and the time at which it occurs. The wind can hold back the tide, or push it along.

QUESTION 2.10   If a head of water 10 m in height exerts a pressure of 1 atmosphere (1 bar), then what is the effect on the local sea-level of a fall in atmospheric pressure of 50 millibars, as might occur when a severe storm passes?

Thus, not only wind changes but changes in atmospheric pressure can cause the actual tide level to be very different from the expected value, especially during storms. The combined effects of wind and low atmospheric pressure can lead to exceptionally high tides, termed **positive storm surges**, which threaten low-lying coastal regions with the prospect of flooding. On the other hand, some areas would experience abnormally low tides, termed **negative storm surges**, which cause problems in shallow seas for large ships such as supertankers which have a relatively deep draught.

The most catastrophic positive surges are those caused by tropical cyclones (typhoons and hurricanes) or by severe depressions in temperate latitudes. The worst in recent history struck the north coast of the Bay of Bengal in 1970, killing 250000 people; and another in 1985 caused the loss of 20000 lives. The well-documented North Sea storm surge of 1953 led to local sea-levels up to 3 m above normal and caused 1800 deaths in Holland and 300 in England. In this case (as with most positive surges), high spring tides, strong onshore winds and very low barometric pressure combined to produce an abnormal rise in local sea-level. In 1986, more than 30 years after this disaster, a barrier 8 km long was built across the eastern Scheldt, completing the final stage of the Delta project which is intended to protect the Netherlands from another flood catastrophe. The completion of the Thames Barrage has provided similar protection for the low-lying areas in and around London. Early warning of storm surges can be given if accurate meteorological and tidal data are available. Forecasting can be aided by satellite tracking of storms, and by computer-modelling of past surges.

Storm surges in the North Sea can, in theory, add up to an extra 4 m to the normal tidal height, but fortunately most storm surges (of which there are, on average, about five per year) are of the order of 0.5 to 1 m. They are usually associated with eastward-moving depressions, and follow a three-phase pattern:

1   The first signs are evident as a relatively small positive storm surge in the North Atlantic, whereby water is displaced by south-westerly winds to the north-east Atlantic.

2   At the same time as the events in (1), a negative surge is experienced on the east coast of Britain as the south-west winds displace water to the north-east corner of the North Sea. This negative surge travels southwards down the east coast and swings eastward across the southern part of the North Sea, following the amphidromic system shown in Figure 2.8.

3   As the depression moves across Britain and out over the North Sea, the wind veers (i.e. swings in a clockwise direction) to blow from the north-west. The next high tide, by now travelling southwards down the North Sea, is thus reinforced not only by the wind but by the Atlantic surge referred to in (1) above, which by this time is displacing water into the northern part of the North Sea. This large positive surge travels down the east coast of Britain, and reaches a maximum in the south-western corner of the North Sea. The problem is compounded, not only by the funnelling effect imposed by the basin shape, but by the fact that the arrival of the surge may coincide with the arrival of the low pressure area in the centre of the depression, thus increasing local sea-level still further.

2.4.3   TIDES IN RIVERS AND ESTUARIES

Many of the world's rivers are in part tidal, because sea-levels have risen since the last glacial period. In such cases, the lower river valleys have become inundated by the sea, forming estuaries or rias. The tides thus propagate up the estuaries, and in some cases into the lower reaches of the rivers. The distinction between an estuary and the tidal reaches of a river is somewhat blurred, and for the purposes of this Section the two are treated as one. The speed of tidal propagation into an estuary depends upon the water depth. Hence the wave crest (high water) will travel faster than the wave trough (low water). As a result, there is an asymmetry in the tidal cycle, with a relatively long time interval between high water and the succeeding low water, and a shorter interval between low water and the next high tide (Figure 2.13).

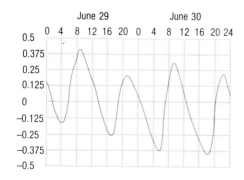

Figure 2.13   A tidal curve for the Hudson River estuary near Albany, New York, showing a typical river tide with peaks tending to catch up with the preceding trough. Numbers on the horizontal axis are time in hours.

The maximum speeds of the tidal currents associated with estuarine tides may not always be in phase with the tidal crests and troughs. Thus, at the estuary mouth, the maximum speed of the flooding tide may coincide with high water, yet further up-river high tide may well occur at the same time as slack water (i.e. zero current). However, the ebb current will invariably persist for longer than the flood, partly as a result of the asymmetry of the tidal cycle already referred to, and partly because the freshwater discharge into the river results in a net seaward discharge of water. Many towns and cities sited near such estuaries rely upon this net seaward flow to carry away sewage, a strategy which is sometimes only a 'mixed' success.

In some tidal rivers, where either the river channel narrows markedly, or the gradient of the river bed steepens, a **tidal bore** may develop. The formation of tidal bores has features in common with the propagation of waves against a counter-current (see Section 1.5.1). The rising tide may force the tidal wave-front to move faster than a shallow-water wave can freely propagate into water of that depth (*cf.* equation 1.5). When this happens, a shock wave is formed, which moves upstream as a rolling wall of water, or tidal bore. It is analogous to the 'sonic boom' that occurs when a pressure disturbance is forced to travel faster than the speed of sound. Most tidal bores are relatively small, of the order of 0.5 m high, but some can be up to ten times that height. The Severn River bore in England is some 1–2 m high, whereas the Amazon bore (called the *pororoca*) reaches about 5 m, and moves upstream at about 12 knots. Other rivers where bores develop include the Colorado, Trent, Elbe, Yangtze and the Petitcodiac, which flows into the Bay of Fundy, notable for its large tidal range (Section 2.4.2).

2.4.4 TIDAL POWER

Power can be generated by holding incoming and outgoing tides behind a dam, using the head of water so produced to drive turbines for electricity generation. The tidal range controls the potential energy available at any locality, and must exceed 5 m for electricity generation to be economic. Suitable localities are limited to those where such a range exists and where dams can be built practicably (Figure 2.14).

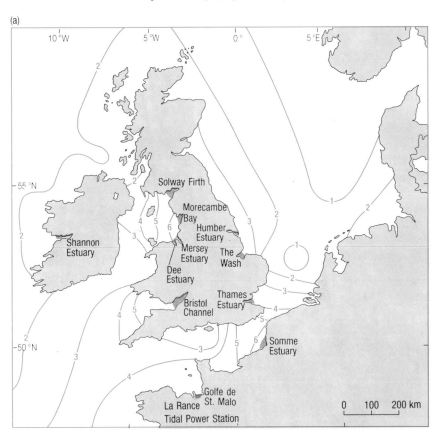

Figure 2.14   The ranges of the M_2 tide (in metres), and the sites of actual (La Rance) and potential tidal-power barrages.

One such site is the Rance estuary in Brittany (Figure 2.15(a) and (b)), which has been in use since 1966. A much larger scheme for Britain's Severn estuary has been proposed and discussed many times. Although such a scheme would produce an appreciable proportion (in the order of 6%) of Britain's electrical power requirements, dam construction would affect the patterns of currents and sediment movements, while ecological disturbance would be inevitable—all factors to be considered whenever schemes of this kind are planned.

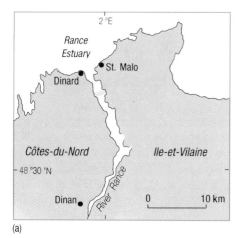

(a)

Figure 2.15(a) The location of La Rance tidal power station. It has been producing about 550×10^6 kWh annually since 1966.

(b) An aerial view of La Rance.

(b)

2.5 SUMMARY OF CHAPTER 2

1   Tides are shallow-water waves, generated by gravitational forces exerted by the Moon and Sun upon the oceans.

2   A centrifugal force, directed away from the Moon, results from the Earth's eccentric rotation around the Earth–Moon centre of mass. This centrifugal force is exactly balanced *in total* by the gravitational force exerted on the Earth by the Moon. However, gravitational force exceeds centrifugal force on the 'moonside' of Earth, resulting in tide-producing forces directed towards the Moon, whereas on the other side of the Earth centrifugal force exceeds gravitational force, resulting in tide-producing forces directed away from the Moon.

3   Tractive forces (horizontal components of tide-producing forces) are maximal on two small circles either side of the Earth, and produce two (theoretical) equilibrium tidal bulges—one directed towards the Moon, and the other directed away from it. As the Earth rotates with respect to the Moon (a period of 24 hours 50 minutes), the equilibrium tidal bulges would need to travel in the opposite direction in order to maintain their positions relative to the Moon. In the simplest case, when the Moon is overhead at the Equator, the tidal bulges travel in the same plane as the Equator, and at all points the two bulges cause two equal high tides daily

(equatorial tides). The Moon is not always overhead at the Equator, but has a declination of up to 28° either side of it, so that the plane of travel of the tidal bulges, when offset with respect to the Equator, gives two unequal, or tropic, tides daily. The declination varies over a 27.2 day cycle. The elliptical orbit of the Moon about the Earth causes variation in the tide-producing forces (up to 20% from the mean value).

4   By analogy with the Moon, the Sun produces equilibrium tides which show inequalities related to the Sun's declination (up to 23° either side of the Equator), and vary in magnitude due to the elliptical orbit of the Earth around the Sun. The Sun's tide-raising force has about 46% of the strength of the Moon's. Solar tides interact with lunar tides. When Sun and Moon are in syzygy, the effect is additive, giving large-ranging spring tides; but when Sun and Moon are in quadrature, tidal ranges are small (neap tides). The full cycle, which includes two neaps and two springs, takes 29.5 days.

5   Tidal speed is limited to about $230 \, ms^{-1}$ in the open oceans (less in shallower seas), and land masses constrain tidal flow. Water masses have inertia, and do not respond instantaneously to tractive forces. Indirect tides, found at low latitudes, lag 90° of longitude behind theoretical equilibrium tides, whereas direct tides occur at higher latitudes, and coincide with the theoretical equilibrium tides. Tides with intermediate characteristics occur at mid-latitudes.

6   The Coriolis force, and constraining effects of land masses, combine to impose amphidromic systems upon tides. High tidal crests circulate around amphidromic points which show no change in tidal level. Tidal range increases with distance from an amphidromic point. A constrained amphidromic system, as in a large bay, tends to rotate in the opposite direction to the deflection imposed by the Coriolis force.

7   The actual tide is made up of many components (partial tides), each corresponding to the period of a particular astronomical motion involving Earth, Sun or Moon. Partial tides can be measured over a long time period at individual locations, and the results used to compute future tides. Actual tides are classified by the ratio of the summed amplitudes of the two main diurnal components to the summed amplitudes of the two main semi-diurnal components.

8   Tidal rise and fall produces corresponding lateral water movements (tidal currents), the speeds and directions of which are influenced by the geometry of the basin and its constraining land masses. Areas of low atmospheric pressure cause elevated sea-levels, whereas high pressure depresses sea-level. A strong wind can hold back a high tide or reinforce it. Storm surges are caused by large changes in atmospheric pressure and the associated strong winds. Positive storm surges often result in catastrophic flooding.

9   In estuaries, the tidal crest travels faster than the tidal trough because speed of propagation depends upon water depth; hence the low-water to high-water interval is shorter than that from high-water to low-water. Tidal bores develop where tides are constrained by narrowing estuaries and the wave-front is forced by the rising tide to travel faster than the depth-determined speed of a shallow-water wave. Where tidal ranges are large and the water can be trapped by dams, the resultant heads of water can be used for hydro-electric power generation.

Now try the following questions to consolidate your understanding of this Chapter.

QUESTION 2.11   Write an expression for the tide-producing force at point P on Figure 2.3, using the terms as defined for equations 2.1, 2.2 and 2.3. It is not essential to try to simplify or approximate the expression.

QUESTION 2.12   Which of the following statements are true?
(a)   'In syzygy' has the same meaning as 'in opposition'.
(b)   Neap tides would be experienced during an eclipse of the Sun.
(c)   Spring tides do not occur in the autumn.
(d)   The lowest sea-levels of the spring–neap cycle occur at low tide while the Moon is in quadrature.

QUESTION 2.13   Briefly summarize the factors accounting for differences between the equilibrium tides and the observed tides.

QUESTION 2.14   How will each of the following influence the tidal range at Immingham (Figure 2.11):
(a)   The Earth's progress from perihelion to aphelion?
(b)   The occurrence of a tropic tide?
(c)   A 30 millibar rise in atmospheric pressure?
(d)   A spring tide?

QUESTION 2.15   The caption to Figure 2.1 states that the view is from above one of the poles. Which pole is in view?

CHAPTER 3   AN INTRODUCTION TO SHALLOW-WATER ENVIRONMENTS AND THEIR SEDIMENTS

'The waves massed themselves, curved their backs and crashed. Up spurted stones and shingle.'
From *The Waves* by Virginia Woolf.

Shallow-water environments are the coastal and shallow marine regions which form the interface between the deep ocean basins and exposed land surfaces. Some of them, such as beaches, tidal flats and estuaries, are very familiar, not least because their natural beauty or their associated leisure opportunities make them popular holiday locations. In the later Chapters of this Volume we examine the patterns of sediment movement and deposition in these environments and see how these patterns are controlled primarily by wave action and tidal currents. In this Chapter we look first at the supply of sediments to the coastal and shallow marine environments, and then consider how this supply has varied over the past few tens to hundreds of thousands of years. Finally, we consider briefly the environments themselves, to see how they are related to each other and how they change over long periods of time.

3.1 THE SUPPLY OF SEDIMENT TO SHALLOW-WATER ENVIRONMENTS

The sediments found in coastal and shallow marine environments around Britain and other mid- to high-latitude countries are derived predominantly from the weathering and erosion of continental rocks. In other words, they are *terrigenous* in origin. The most important solid products of weathering are rock fragments, quartz grains and clay minerals. Quartz is the dominant mineral in the sand deposits that form beaches in the mid-latitudes, and clay minerals are the most important constituents of the muds of tidal flats and estuaries. Sands and muds are also widespread as nearshore and coastal sediments in many tropical and sub-tropical areas. Clay minerals are very fine-grained (less than 2μm in grain diameter) flaky minerals which stick together easily and give mud its delightfully glutinous properties.

Along arid coastlines and other areas, where the supply of terrigenous sediment is limited or absent, quartz beach sands are replaced by carbonate sands. These are made up of shell material composed of calcium carbonate extracted biologically from seawater, and even some grains composed of calcium carbonate which has been chemically precipitated from seawater. Similarly, the clay-rich muds of tidal flats and estuaries are replaced by equally fine-grained carbonate muds.
The products of weathering are carried to the ocean margins by rivers. The main dissolved products of weathering, found as constituents of river water, are calcium, sodium, potassium and magnesium cations and bicarbonate anions. Most of the solid material is carried within the water flow as suspended sediment, but some is rolled or bounced along the river bed as **bedload**. The supply of sediment is by no means evenly distributed around the ocean margins, as Figure 3.1 demonstrates.

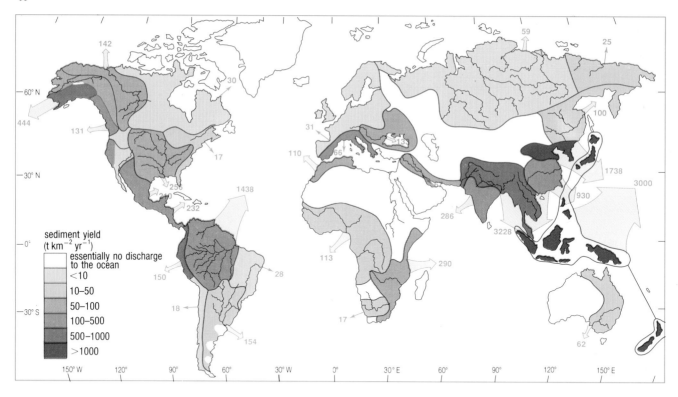

Figure 3.1 The average annual discharge of
suspended sediment from the major drainage basins
of the world. Values are 10^6 tonnes yr^{-1}. The
sediment yields of the various basins are shown in
the key, and the sediment discharge is proportional
to the widths of the arrows. Red lines show
boundaries of the main drainage basins. Note that the
sediment loads of the Nile and Colorado rivers are
largely trapped in reservoirs behind dams, and so do
not reach the sea.

QUESTION 3.1 Examine Figure 3.1 and, with the aid of an atlas and the
ocean floor map, explain the following.

(a) Why is more sediment discharged into the Atlantic Ocean from
Brazil than is discharged into the Pacific Ocean from Ecuador and Peru,
even though extensive weathering occurs along the Andean Cordillera,
which runs down the Pacific margin of South America?

(b) Why is sediment discharge so high in the south-west Pacific Ocean?

(c) Why does so little sediment reach the deep south-west Pacific Ocean
basin compared with the deep Atlantic Ocean basin?

Despite their immense importance, rivers are not the only means of
transporting sediment to the oceans, as Figure 3.2 shows. Ice, wind,
coastal erosion and volcanic eruption each play a role that varies in
importance from one part of the world to another. Sources of ice-borne
sediment are generally restricted to high-latitude regions where glaciers
and ice-sheets enter the sea. By contrast, wind-blown dusts are found
mainly at low latitudes where the Trade Winds blow off continental
desert areas.

On average, volcanic debris makes a small contribution and volcanic
eruptions are localized in their occurrence. However, following a volcanic
eruption, the input of volcanic debris may be locally overwhelming. For
example, in 1815 the Indonesian volcano of Tambora erupted around
40×10^9 tonnes of volcanic material, much of which must have been
deposited eventually in the oceans. Volcanic material ejected high up in
to the atmosphere is, of course, dispersed by winds. The speed with which
such dusts can sink into quite deep ocean waters was revealed when
radioactive contaminants ejected as a result of the Chernobyl nuclear
power station disaster in 1986 were identified at depths of several hundred
metres in the Mediterranean Sea a few weeks later.

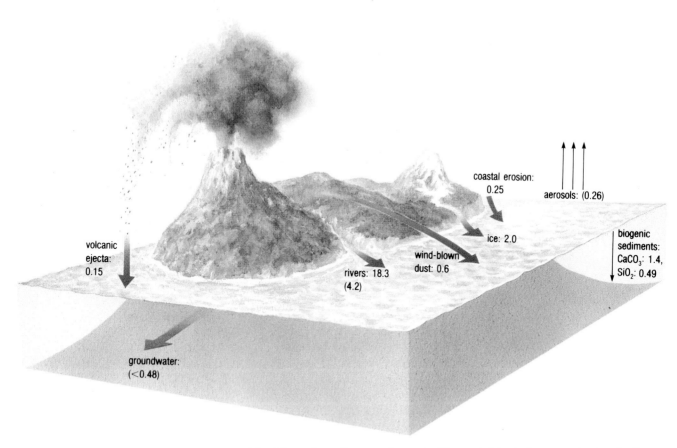

Figure 3.2   The annual transfer of sedimentary materials to the oceans in 10^9 tonnes per year. Numbers in brackets refer to material in solution.

The biogenic sediments shown in Figure 3.2 include not only the shallow-water carbonates mentioned earlier, but also the deep-water carbonate and siliceous sediments deposited in the deep ocean basins.

To what extent do you think the amounts given in Figure 3.2 are reliable?

Such amounts can only be estimates based on measurements made by various researchers over relatively brief periods of time at a limited number of localities. The techniques used to make measurements, the length of time over which the measurements are made and the sampling procedures also vary greatly from one researcher to the next, and, as new data are collected, these estimates are constantly updated. Nevertheless, Figure 3.2 does provide an adequate indication of the *relative* importance of different sediment sources.

3.2   VARIATIONS IN THE SUPPLY AND DISTRIBUTION OF SHALLOW-WATER SEDIMENTS

During the past two million years, the most important changes which have affected present-day shallow-water sediments have been those resulting from the climatic fluctuations associated with the Quaternary glaciations. These changes caused variations in the supply of glacial and river sediments. However, the related changes in sea-level which occurred as ice-sheets grew and declined were more important, as they affected the types of sediment that are found on the continental shelves today.

As the ice-sheets grew, water was withdrawn from the oceans and the sea-level fell. Consequently, areas which were previously (and are now) continental shelf, covered by shallow seas, became exposed above sea-level as land. For example, during the most recent sea-level minimum, about 80000 to about 20000 years ago, the sea-level dropped to some 110–120m below its present level. This meant that rivers and glaciers extended much further seawards and so river and glacial sediments were deposited on what is now the submarine continental shelf. When the ice melted, and the seas inundated the shelves once more, these terrestrial deposits remained on the shelves, below sea-level, as **relict sediments** where they are now being reworked by storm waves and tidal currents.

3.3 THE RELATIONSHIPS BETWEEN SHALLOW-WATER ENVIRONMENTS AND THEIR CHANGES WITH TIME

In the following Chapters we shall consider, in turn, beaches, tidal flats, estuaries, deltas and shelf seas. Although these environments are treated as separate topics, it is important to note that such divisions are artificial. As you will see, whether a beach rather than a tidal flat develops is a question of whether wave action at the coastline has more effect on sediment transport than have tidal currents. Many stretches of coastline are influenced more or less equally by both waves and tidal currents and so show features characteristic of both beaches and tidal flats. The change from beach to shelf sea is defined arbitrarily as the depth of water below which sediment is not usually disturbed by waves during normal fair weather conditions. Estuaries are river mouths into which the sea penetrates at high tide. They are very much influenced by tidal currents and so the pattern of sediment deposition shows many similarities to that of tidal flats. However, as increasing amounts of sediment are brought down by rivers, these sediments may build seawards at the river mouth to form a delta. The sands of many deltas are in turn reworked by waves to form beaches.

Shallow-water environments are transient, existing only for relatively brief periods of geological time. For example, marine erosion is cutting back the cliffs of parts of the Norfolk coast at rates measured in metres per year whereas marine deposition is extending the Dungeness foreland at a similar rate. The lifespan of a lobe of the Mississippi delta, from initial accretion of the sediments to their eventual subsidence, is estimated to be around a thousand years. Most present-day estuaries originated when sea-levels rose close to their present levels, following the end of the last major glaciation, about 18000 years ago. Sea-level began to stabilize about 6000 to 10000 years ago and ever since then estuaries have begun to silt up quite rapidly. As you know, even the shallow seas over the continental shelves are not permanent features; they advance and retreat on the same time-scales as global changes in sea-level. Today's familiar coastal features around Britain are not the same features that the Romans encountered when they invaded Britain some 2000 years ago, nor are they the same as those that will be appreciated by the generation that celebrates the year 4000.

3.4 SUMMARY OF CHAPTER 3

1   The sediments along continental margins in mid- to high latitudes are predominantly terrigenous in origin and consist of rock fragments, quartz sands and clay-rich muds. Sands and muds are also widespread at low latitudes. Along low-latitude arid coastlines, and other regions where terrigenous sediments are scarce or absent, quartz sands are replaced by carbonate sands, and clay-rich muds are replaced by carbonate muds.

2   River transport is the most important means of bringing terrigenous sediments to the ocean margins, but ice transport, wind transport and volcanic eruptions may be locally important.

3   Periodic falls in sea-level during the Quaternary have resulted in the deposition of river and glacial sediments on areas of the continental shelf which are today covered by the sea. These relict sediments are now being reworked by waves and tidal currents.

4   Divisions between different types of shallow-water environment are not clear cut, and one type of environment passes transitionally into another. These environments change gradually from one type to another on time-scales of hundreds up to thousands of years.

Now try the following question to consolidate your understanding of this Chapter.

QUESTION 3.2   Which of the following types of shallow marine sediments tend to be latitudinally restricted in their distributions?

(a)   Relict sediments.

(b)   Glacial sediments.

(c)   Wind-blown dusts.

(d)   Volcanic sediments.

(e)   Carbonate sediments.

| CHAPTER 4 | SEDIMENT MOVEMENT BY WAVES AND CURRENTS |

'Listen! you hear the grating roar
Of pebbles which the waves suck back, and fling,
At their return, up the high strand.'

From *Dover Beach* by Matthew Arnold.

One important consequence of wave and tidal current movements in coastal and nearshore waters—as you should know if you have ever stood on a beach listening to the waves—is that some of their energy is transferred to the movement of sediment. What sort of sediment is moved, how much, and where it is moved to, depend on factors such as the energy of the waves and tidal currents and their direction of motion. Over relatively short periods of time, there is a natural equilibrium between the amount and rate at which sediment is supplied to the nearshore region, and the redistribution of this sediment by water movement. This equilibrium must be properly understood by marine engineers when they build jetties and breakwaters to protect harbours, construct groynes to prevent beach erosion or lay submarine cables and pipelines. Otherwise, the equilibrium may be disrupted with disastrous consequences either for the constructions themselves, or for adjacent stretches of coastline, and sometimes for both.

In this Chapter we consider, in general terms, the physical conditions which lead to the movement and deposition of sediment before going on to discuss the effects of waves and tides on sediments in different shallow-water environments. The theory explained in this Chapter may seem complex, but it is necessary if we are to predict and quantify the sediment movements that are of increasing importance to marine engineers. The full theory is actually a good deal more complex than that provided in this Chapter. And, as you will see later, certain modifications are necessary when the theory is applied to the marine environment because of the interaction of tidal currents and waves. However, you will find that much of the theory is relevant to processes discussed in more detail in later Chapters.

4.1 FLUID FLOW

From your own observations, you will realize that the gentle waves which break on to a flat sandy beach are capable of washing sand grains up and down the beach but could not normally shift pebbles. In order to make sediment move, the force of the water flowing over it must be capable of overcoming both the force of gravity acting on the sediment grains and the friction between the grains and the surface on which they are resting. This fluid force is composed of a buoyancy component, acting vertically upwards (lift component) and a frictional force between the flowing water and the underlying grains (drag component) (see Figure 4.1). Once a sediment grain has begun moving, the influence of the lift component dies away rapidly and movement is influenced increasingly by the frictional force.

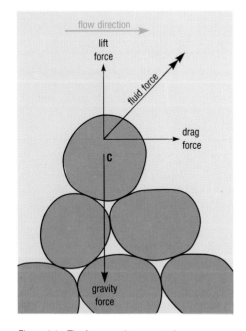

Figure 4.1 The forces acting on a stationary sediment grain resting on a bed of similar grains beneath a flow. Point C is the centre of gravity of the grain. The drag and lift forces do not necessarily act through the centre of gravity because the flow around the grain will be influenced by the shapes of the surrounding grain surfaces.

4.1.1 FRICTIONAL FORCES AND THE BOUNDARY LAYER

When water flows over sediment lying on the sea-bed or a river bed, there is frictional drag between the water and the sediment. The influence of the friction extends upwards some way towards the surface of the water. The effect of this friction is to slow down the water flow so that the actual speed of the water you might observe at the surface by dropping a twig on to the water is much greater than that occurring near the bed. The lower part of the flow which experiences frictional retardation is known as the **boundary layer**. In the sea, the boundary layer normally extends between 1m and 10m above the sea-bed, but in shallow water it may well occupy the whole water depth. A boundary layer develops wherever a fluid moves over a surface, whether it be water over the sea-bed, winds over the sea-surface, or syrup over a table top.

If you were able to measure the speed of water at various depths in the water column, from the bed up to the surface, you would find that it varies systematically. Provided no sediment on the bed is moving, theoretically the imperceptibly thin layer of water in direct contact with the bed is also stationary and so the speed should be zero. However, the layer of water immediately above this does move, albeit very slowly, and so slides, or shears over the lower layer. With increasing distance from the bed, successive layers of water move a little faster as the effects of frictional retardation by the bed decrease. Each successive layer therefore shears over the layer beneath. Thus, the frictional force that is responsible for sediment movement is a **shearing force**. As you move up the water column, a shearing force also operates on each layer of water as a result of the increase in relative speed of the layer above. However, the *rate* at which speed increases gradually lessens with increasing distance from the bed as the influence of frictional retardation by the bed begins to die out. Eventually, the speed stops increasing and reaches a constant value. The height above the bed at which this occurs represents the top of the boundary layer.

Within the boundary layer, then, there is a velocity gradient, and a graph of height above the bed against velocity of the flow is the **velocity profile** for the flow (Figure 4.2).

Because of the shearing force caused by layers of water moving over each other, a stress is created on each layer which is called a **shear stress**, designated τ (tau—pronounced 'taw'). The value of the shear stress operating at the actual bed (τ_0) is crucial when we consider whether or not sediment on the bed is likely to be moved. You have already met the concept of shear stress in the guise of frictional stress in Sections 1.1.2 and 2.4.1. The frictional stress generated when wind (a moving fluid) passes over the ocean surface (a fluid moving more slowly) is a shear stress. In that case, the transfer of energy results partly in the formation and movement of waves; when water moves over sediment on the sea-bed, or a river bed, the transfer of energy results in the movement of sediment and possibly the formation of sediment ripples.

QUESTION 4.1   What must the units of τ_0 be for water flowing over an area of the sea-bed?

The value of τ_0 is related directly to the rate at which the velocity (u) of moving water increases with height (z) above the bed. So, if we express

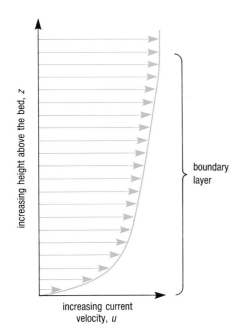

Figure 4.2 The velocity profile for a steady current flow over a bed. The arrows indicate the direction of flow and the length of each arrow is proportional to the current velocity at that height above the bed.

74

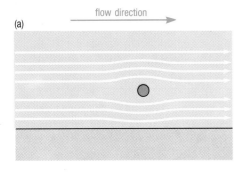

(a) flow direction

(b) mean flow direction

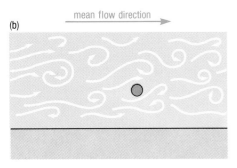

Figure 4.3(a) Laminar flow in a fluid: the molecular flow of the fluid round a grain is smooth.

(b) Turbulent flow in a fluid: complex, multidirectional eddy currents are superimposed on the overall flow direction.

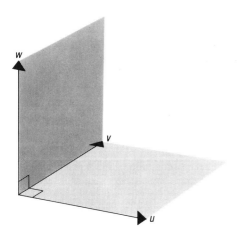

Figure 4.4 The three component velocities of fluid particles in turbulent eddies. The *u*-direction is horizontal, parallel to the flow direction; the *v*-direction is horizontal and at 90° to the flow direction and the *w*-direction is vertical.

the change of velocity as du, and the change in height above the bed as dz then we can say:

$$\tau_0 \propto \frac{du}{dz} \tag{4.1}$$

The faster a body of water moves, the greater is the rate of change of velocity with height above the bed, and so the shear stress increases. There is, however, another factor which affects the value of τ_0, and that is the **molecular viscosity** (μ—pronounced 'mu') of the water. This is simply a measure of the ease with which a fluid will flow and is a constant physical property for any fluid at a given temperature. Glycerol and treacle, which are very thick fluids at room temperature, do not flow easily and so their viscosities are much higher than that of water. Taking viscosity into account, we can now amend equation 4.1 to:

$$\tau_0 = \mu \times \frac{du}{dz} \tag{4.2}$$

4.1.2   THE FLOW OF WATER IN THE BOUNDARY LAYER

When water flows slowly over a flat surface, all the water particles passing a given point will follow exactly the same path—a stream line. This condition is known as **laminar flow**. It means that if a sediment particle is caught in the flow, the stream lines will flow smoothly round it and resume their former path on the other side (Figure 4.3(a)). As the flow becomes faster, the stream lines begin to break down and the water particles start to move in random eddies throughout the fluid, although the net direction of movement is the direction of the main flow (Figure 4.3(b)). This is now said to be **turbulent flow** and the turbulent eddies lead to an *effective* viscosity (or **eddy viscosity**, η—eta, as in 'cheetah') which can be up to several orders of magnitude larger than the molecular viscosity.

Flows in the sea and all natural water flows are almost always turbulent.

QUESTION 4.2(a)   Examine the velocity profile shown in Figure 4.2. How do you think the flow conditions will vary with increasing distance from the bed through the boundary layer?

(b)   In the light of your answer to (a), how do you think the viscosity will change with height above the bed?

Because of the development of eddies in turbulent flow, water particles are able to move in all directions about the net direction of flow. This means that the horizontal flow velocity can be broken down into three velocity components acting at right angles to each other (Figure 4.4): the u-component is horizontal, and parallel to the net flow direction; the v-component is horizontal but at right angles to the net flow direction, and the w-component is vertical.

The turbulent shear stress is much greater than the shear stress that exists for laminar flow and has been shown, experimentally, to be proportional to the square of the mean velocity $\bar{u}$:

$$\tau_0 \propto \bar{u}^2 \tag{4.3}$$

This is a fundamentally very important relationship because all flows in the oceans are turbulent. In equation 4.3, you should have noted that the term for velocity, u, is shown with a bar above it, $\bar{u}$. This represents the mean horizontal velocity averaged over a period of time, taking into account the fluctuations in water particle directions caused by turbulent eddies. For turbulent flow, equation 4.2 then becomes:

$$\tau_0 = (\mu + \eta) \times \frac{d\bar{u}}{dz} \tag{4.4}$$

When water flows over a relatively smooth, flat bed (a condition known as **smooth turbulent flow**) the velocity profile of the lower part of the boundary layer is like that shown in Figure 4.5(a). Directly next to the bed there is a very thin layer of water (in the order of a few millimetres thick) where conditions approach laminar flow and the rate at which velocity increases with height above the bed can be considered as constant for all practical purposes.

As du/dz is virtually constant within this layer, and the viscosity is dominantly the molecular viscosity of the water, what can you say about the value of τ in this layer?

The value of τ must be almost constant too, because it is proportional to both du/dz (which is virtually constant) and the molecular viscosity, which is also a constant at a given temperature. Therefore, this part of the velocity profile is almost linear and appears as a straight line. As the molecular viscosity is important to the value of τ in this layer, the layer is known as the **viscous sublayer**.

Above the viscous layer and below the overlying fully turbulent layer is a transitional, or buffer, layer. In the lower part of this layer, the rate of increase of velocity decreases logarithmically with increasing distance from the bed, and so this part of the velocity profile is curved. The value of τ is not constant in the turbulent layer because the rate of change with height ($d\bar{u}/dz$) is changing, and the value for the eddy viscosity is changing, too.

4.1.3   THE SIGNIFICANCE OF THE VISCOUS SUBLAYER

You might consider that as the viscous sublayer is at most only a few millimetres thick, it is not worth bothering about. However, its presence affects the movement of sediment quite considerably. To begin with, it is not a constant thickness—its thickness varies according to the current speed.

QUESTION 4.3   Would you expect the thickness of the viscous sublayer to increase or decrease as current speed increases?

Another factor that affects the presence of the viscous sublayer is the grain size of the sediment on the bed in relation to the thickness of the sublayer. When the grain diameter (and here we are referring to the most abundant grain size within the sediment, rather than the occasional larger grains) is less than one-third of the thickness of the viscous sublayer, the sublayer remains intact. The main body of the turbulent flow above is 'unaware' that these grains exist and so does not readily move them (Figure 4.5(a)).

If the current flows over sediments with grain diameter greater than about one-third of the thickness of the viscous sublayer, the grains begin to

76

(a)

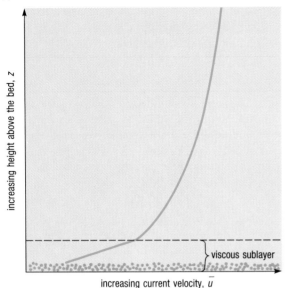

(b)

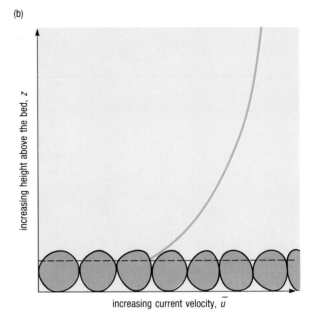

Figure 4.5(a) A velocity profile for smooth turbulent flow over a fine-grained bed. The profile is virtually linear within the viscous sublayer which develops close to the bed.

(b) A velocity profile for rough turbulent flow over a coarse-grained bed. The viscous sublayer (potential thickness is indicated by the dashed line) is disrupted by the coarse grains. Turbulent flow (the logarithmic layer) extends very close to the bed.

disrupt the flow in the sublayer so that the smooth turbulent flow conditions begin to break down (see Figure 4.5(b)). Once grain diameters have reached about seven times the expected thickness of the sublayer, they protrude so far into the turbulent layer that the sublayer breaks down altogether and turbulent conditions extend right down to the bed. It is now described as **rough turbulent flow**. Turbulent eddies are now able to reach down between the sediment grains and so there is much greater potential for sediment movement.

The conditions under which either smooth or rough turbulent flow will occur depend on both the flow speed and the grain size of the underlying sediment.

QUESTION 4.4 For each of the flow conditions outlined in (a) and (b) below, explain how you would expect flow roughness to change and how the change might affect the potential for sediment movement.

(a) A current of constant speed passes from a very fine-grained bed to a coarse-grained bed.

(b) A current flowing over a fine-grained bed increases in speed.

4.1.4 CURRENT SHEAR AT THE BED

From Section 4.1.3, it should be clear that changes in speed, and therefore changes in the velocity profile of a flow, have a marked effect on sediment movement at the bed. If we were able to estimate the shear stress operating at the bed, we might also be able to estimate what sort of grain size was present on the bed, and what size of grain was likely to be moved by flowing water. The shear stress at the bed (τ_0) is related to the rate of change of velocity over the whole profile. This can be determined relatively easily by suspending electromagnetic current flow meters within the flow at various depths above the bed and plotting the results on a graph. However, because the velocity profile in the boundary layer takes the form of a curve (Figure 4.6(a)), calculation of $d\bar{u}/dz$ would involve some quite complex mathematics. There is a way of overcoming this.

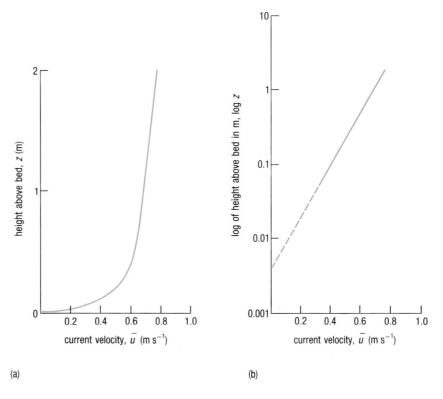

Figure 4.6(a) The velocity profile for a water flow over a bed plotted using a linear scale on both the vertical and horizontal axes.

(b) The same velocity distribution as in (a) plotted using a $\log_{10}$ vertical scale and a linear horizontal scale.

(a)

(b)

Recall that in the lower part of the turbulent boundary layer the rate of increase with height above a bed decreases logarithmically. This means that if we plot the current meter data on a graph using a log scale for height above the bed, then the velocity profile appears as a straight line (Figure 4.6(b)). The reason the line in Figure 4.6(b) is dashed below about 0.06 m (6 cm) from the bed is that below this depth it becomes impractical to place current meters to obtain further readings. The dashed line is therefore an extrapolation on to the depth ($\log z$) axis and takes no account of the viscous sublayer. (The reason why the graph intercepts this axis at a finite distance above the sea-bed will be explained later (Section 4.1.5).)

We are now in a position to measure $d\bar{u}/dz$ directly from the graph. This is referred to as the 'velocity gradient' (Section 4.1.1) although you will realize that it is actually the *inverse* of the gradient of the graph which is being measured. There is a good reason for this. Conventionally, when plotting a graph, the variable that can be fixed is plotted on the *x*-axis (or horizontal axis) and the variable that is to be measured is plotted on the *y*-axis (or vertical axis). Thus, it would be more usual to plot height above the bed on the horizontal axis because the positions of the current meters can be fixed, and velocity on the vertical axis because this is what will be measured. In this case, $d\bar{u}/dz$ *would* be the gradient of the graph. However, the visual effect of a change of velocity with height above the bed, and the velocity profile, would be lost, and so when constructing a velocity profile it is normal to reverse the axes of the graph.

At this point, there is a further convention to observe. It is actually quite difficult to grasp the meaning of a stress measured in $N\,m^{-2}$ in relation to a body of water which we see flowing at a speed measured in $m\,s^{-1}$. Therefore, it is usual to convert shear stress into a term that has the units of velocity. This term is designated u_* and is called the **shear velocity** or friction velocity. It is a purely fictitious quantity (in other words, it cannot be measured by suspending current meters in a water flow). The shear velocity of a fluid is derived mathematically from the shear stress (τ_0) and the density of the fluid (ρ — rho) by the relationship:

$$u_* = \sqrt{\frac{\tau_0}{\rho}} \tag{4.5}$$

or $\quad \tau_0 = \rho u_*^2$ (4.6)

Note once again that the shear stress is proportional to the *square* of a velocity term.

QUESTION 4.5   The newton is a unit of force and so has units that are mass × acceleration = $kg\,m\,s^{-2}$. Using this information, show how u_* has units of velocity.

As u_* is directly related to the shear stress, it can also be calculated using the velocity gradient, $d\bar{u}/dz$.

There is one other point to bear in mind before we attempt to calculate u_* from a velocity profile. The actual velocity at any given depth in a flow is related to the velocity gradient by a constant. When the velocity gradient is plotted using $\log_{10}$ on the depth scale, this constant has the value 5.75. It is known as the von Karman constant, (κ—'kappa') after the man who derived it. Using this constant, u_* is related to the velocity gradient as follows:

$$5.75\, u_* = \frac{d\bar{u}}{d\log z} \tag{4.7}$$

By rearranging this equation, we can find the value of u_*:

$$u_* = \frac{d\bar{u}}{5.75\, d\log z} \tag{4.8}$$

An example of how this calculation is made using a velocity profile is shown in Figure 4.7.

QUESTION 4.6(a)   Using the value of u_* calculated from Figure 4.7, work out the value for the shear stress at the bed (τ_0) beneath the flow. (Assume the density of water is $1000\,kg\,m^{-3}$.)

(b)   How does the value of u_* compare with the actual time-averaged current speed 1 m above the bed?

(c)   Is the shear velocity calculated using the depth interval 1 m to 10^{-1} m the same as that between 10^{-1} m and 10^{-2} m on Figure 4.7?

4.1.5   ROUGHNESS LENGTH

You will recall that the velocity profiles shown in Figures 4.6(b) and 4.7 intersect the depth axis at a finite distance above the bed (z_0) implying that zero velocity is reached at a finite distance above the bed. This paradox is resolved quite easily if you remember that the data from which these figures were constructed represent current velocities only for the logarithmic part of the boundary layer, which plots as a straight line on a graph where the depth scale is logarithmic. This is because the lowest

Figure 4.7 The calculation of shear velocity (u_*) from a velocity profile with $\log_{10}$ on the vertical scale. The significance of z_0, the roughness length, is explained in Section 4.1.5.

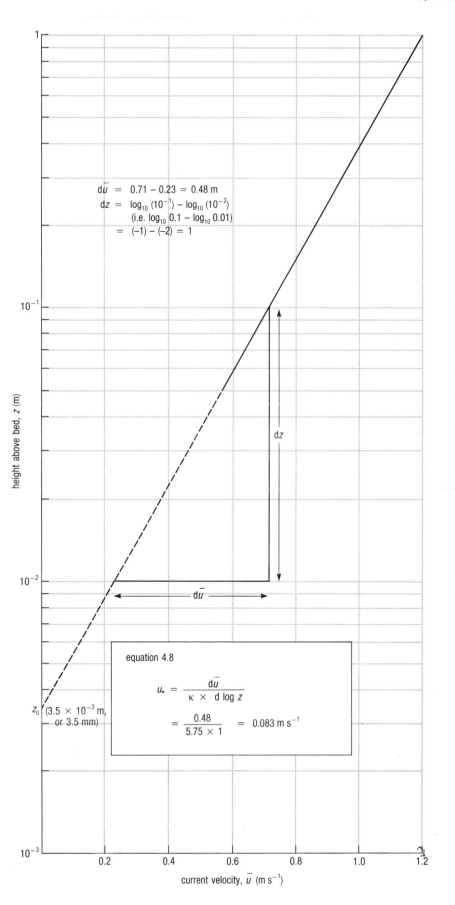

80

current meters are seldom lower than a few centimetres from the bed. When flow conditions are not fully rough, the velocity profile passes through the transitional zone and the viscous sublayer (Figure 4.5(b)). Velocity changes in the viscous sublayer plot as a straight line on Figure 4.6(a), so they would plot as a curve on Figures 4.6(b) and 4.7. When conditions are fully rough, grains break through the viscous sublayer (Figure 4.5(b)). This means that in both sorts of conditions the intercept, z_0, is merely the extrapolation of the logarithmic profile on to the depth axis, and it is known as the **roughness length**. The importance of this value is that it is related both to the grain size of the sediment on the bed, and to upraised features such as ripple marks, where these occur. Its value increases as the grain size of the sediment on the bed, or the size of ripples in the sediment increase, and so the velocity profiles constructed from current meter data can also provide an idea of the roughness of the bed beneath the flow.

QUESTION 4.7(a)   How would you expect the slope of the graph in Figure 4.6(b) to change if:

(i)   the current increased in speed, but continued to flow over a bed with the same roughness?

(ii)   the current remained steady, but flowed over a bed with much coarser grain size?

(b)   What are the implications of these changes for the value of τ_0 and erosion at the bed?

4.1.6   VELOCITY PROFILES IN THE SEA

There are a number of reasons why marine logarithmic velocity profiles are not necessarily straight lines, and therefore why the shear stress estimated from the slope of a profile may not be the shear stress responsible for moving sediment at the bed. These factors have to be borne in mind when interpreting marine velocity profiles.

To begin with, currents in the sea are not always in the same direction; tidal currents, for example, change direction with time as shown in Figure 2.12(a). Also, they accelerate from what may be effectively zero speed at slack water towards a maximum speed and then decelerate again (Section 2.4.1 and Figure 2.12(b)). The result is that the logarithmic profile is curved, as shown in Figure 4.8. The shear stress at the bed and values of both u_* and z_0 can be underestimated for accelerating currents, and overestimated for decelerating currents. However, for most situations in the sea, tidal current acceleration and deceleration occurs near to slack water when the potential for sediment movement is low anyway.

Secondly, turbulent boundary layers have long 'memories', so when a current flowing over the sea-bed encounters a surface with a different roughness, it takes time for the velocity profile to adjust. When the bed roughness is greatly increased due to ripple marks, the ripples form a physical obstruction to the flow which produces a pressure gradient above the ripple, causing an added resistance to the flow. This resistance is called the **form drag**. It reduces the capacity of the flow to move sediment rather than increasing it, and means that not all the shear stress calculated from the logarithmic velocity profile is available to move sediment. For tidal current flow, about half the shear stress is available for sediment

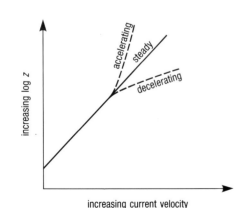
Figure 4.8   Curvature of the log velocity profile produced by accelerating and decelerating flows, e.g. tidal currents.

movement, and half is caused by form drag. When water movement is due to the oscillatory motion of waves (see Figure 1.8(c)), then as little as 10% of the shear stress may be available for sediment movement.

Thirdly, we have been assuming that current flows in the sea consist only of water. As these flows are capable of transporting sediment close to the bed (where most bedload transport occurs), they are a quite concentrated mixture of sediment and water. The sediment concentration, and hence the density of the mixture, decreases away from the bed and so there is a density gradient. It is more difficult for turbulent eddies to move the denser fluid upwards and so the density gradient will tend to dampen out the turbulence and lead to a lower than expected shear stress at the bed.

Finally, it is important to remember that water movements in shallow coastal waters involve more than tidal currents. The picture is complicated by wave action and other types of water movement along the shoreline. We shall consider the effects of these in Chapter 5.

4.2 SEDIMENT EROSION

To those who work with sediments, the terms mud, clay, silt, sand or gravel have specific grain-size limits, and so you will find it useful to refer to Table 4.1 as you read through this Section.

The movement of sediment on the sea-bed begins when the shear stress (τ_0) becomes sufficiently great to overcome the frictional and gravitational forces holding the grains on the bed (Figure 4.1). This value is known as the **critical shear stress**, designated τ_c, and can be determined for any given grain size. It follows that there will also be a **critical shear velocity** (u_{*c}) which determines sediment movement (see equations 4.5 and 4.6).

The relationship between grain size and critical shear stress is not a straightforward, linear one. To begin with, some sediments are **cohesive** in character and some are not, and cohesiveness has a significant effect on sediment erosion. This cohesiveness results mainly from the presence of clay minerals in a sediment. However, biologically produced films may also bind the surfaces of grains together. Clay particles are flaky and very small, less than 2μm in size. Both in sediments and in suspension in water they tend to form aggregates in which the individual flakes are held together by a combination of electrostatic attraction and the surface tension of the films of water surrounding the flakes. These forces are strong and give muds their glutinous property. Clays increase the overall cohesion of the bed, even when they constitute only a small proportion of the total sediment; cohesion begins to be significant when sediment contains more than about 5–10% of clay by weight. Thus, the stickiness of many predominantly silty sediments may be caused by only a relatively small amount of clay.

Non-cohesive sediments contain coarser sediment grains which are often more equidimensional in shape than cohesive sediments. They lack the physico-chemical interactions that exist between clay particles, and so are free to move independently. Non-cohesive sediments include the carbonate sands of waters where the deposition of terrigenous sediment is negligible, as well as the quartz silts and sands characteristic of shallow marine sediments in mid- to high latitudes. As most of the work

Table 4.1 The classification of sedimentary particles according to size (based on the Wentworth scale).

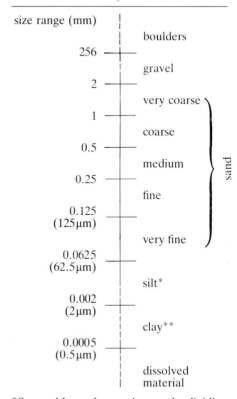

* Some older scales use 4μm as the dividing line between silt and clay on this scale. 2μm is probably the better limit to use because particles <2μm in size tend to be clay minerals, and larger natural particles composed of clay minerals easily break down into particles <2μm in size.

** Muddy sediments typically consist of clay with a variable silt content.

quantifying the movement of non-cohesive sediments has been carried out using quartz grains, we shall not consider carbonate sediments further in this Section.

4.2.1   EROSION OF NON-COHESIVE SEDIMENTS

Consider what happens when the critical shear velocity is reached for the movement of a non-cohesive sediment consisting of, say, equidimensional quartz grains (Figure 4.9). When the grain size is very small, or the flow is very slow, the sediment grains are protected by the viscous sublayer and so no movement occurs. As the shear velocity increases with increasing flow speed, the viscous sublayer starts to break down and the coarsest grains begin to roll or occasionally to slide or bounce across the bed as bedload. This is the area on Figure 4.9 labelled 'transport as bedload'. Eventually, the shear velocity may be sufficient to alternately lift the grains into suspension and temporarily redeposit them, as the turbulent eddies fluctuate. Redeposition occurs when the force of gravity acting on the grains is sufficient to overcome the fluid forces causing movement. The rate at which a grain settles out of suspension, back to the bed, is called its **settling velocity**, which is termed w_s. At even higher shear velocities, the grains are lifted permanently into suspension to be transported as suspended load.

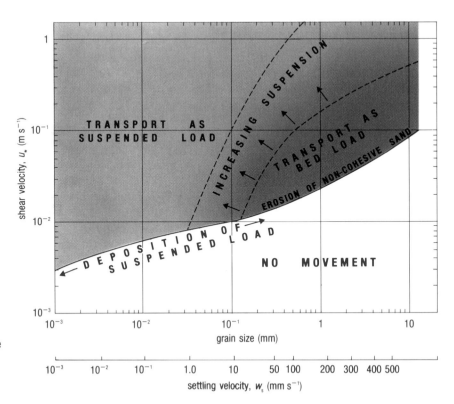

Figure 4.9 Experimentally determined graph showing shear velocity plotted against grain size. The graph summarizes the conditions for the erosion and deposition of sediments of various grain sizes. The settling velocities are given in relation to grain size.

Despite the sharp lines drawn on Figure 4.9, in reality there is a continuous gradation between transport as bedload and transport in suspension.

QUESTION 4.8(a)   Using Figure 4.9, estimate the ratio of the settling velocity to shear velocity (w_s/u_*) that is required to lift grains coarser than about 0.1 mm into suspension initially. From Table 4.1, state what type of sediment this includes.

(b)   Similarly, estimate the ratio of w_s/u_* that is required to lift these grains permanently into suspension.

(c)   Explain what happens to grains finer than about 0.1 mm when the critical shear velocity is reached. Do either of the ratios you have estimated in (a) and (b) apply to these grains?

The fact that fine sands and silts do not go through a bedload phase of transport is important when we consider how sediment ripples form in Section 4.5.2.

You have probably noticed that the scale for settling velocities changes for grains of about 0.1 mm diameter, and coarser. This is because Stokes' law no longer applies to these grains. According to Stokes' law, for grains finer than 0.1 mm the settling velocity is proportional to the square of the grain diameter, $w_s \propto D^2$, where D is the diameter of the grain and the grain is a perfect sphere in an infinite volume of fluid. Above a diameter of about 2 mm, the settling velocity is proportional to the square root of the diameter ($w_s \propto D^{1/2}$), which means that even quite a large increase in diameter leads to only a rather small increase in settling velocity. Over the range 0.1 to 2 mm, the settling velocity is proportional to progressively decreasing powers of the diameter, i.e. from D^2 to $D^{1/2}$.

4.2.2   EROSION OF COHESIVE SEDIMENTS

The strong binding forces that hold cohesive grains together once they have been deposited means the grains cannot be eroded in the same way as can non-cohesive sediments. Cohesive grains are lifted as clumps or 'floccs', rather than as individual grains, and if the muds have become partially consolidated (e.g. on exposed estuarine or tidal mud-flats), erosion occurs following the mass failure of the sediment surface which is ripped off in large lumps—a process that requires very high shear velocities. Thus, once deposited, muds are not easily eroded, despite their fine grain size.

The resistance of mud to erosion is usually assessed by its **yield strength**, which is the maximum shear stress that the sediment can withstand before failure occurs. One method of measuring yield strength is by using a shear vane. This is a device that has a tip comprising two plates at right angles to the end of a shaft. The tip is inserted into a mud sample and force is applied to rotate the shaft. The yield strength is calculated from the force measured at the moment when the sediment yields and the blades suddenly start to rotate. It requires shear velocities equivalent to those needed to move fine gravel in order to erode cohesive, but relatively uncompacted, muds. However, as the degree of compaction increases, very much higher shear velocities are needed to induce sediment failure. This is an important point to bear in mind because it has a considerable effect on the sedimentation in estuaries and tidal mud-flats where muds are particularly abundant. The cohesion of very fine-grained sediments is also influenced by their water content, their mineral composition (e.g. whether they are calcareous or clay-rich), and the salinity of both the overlying water and the water trapped between the sediment grains.

4.3 THE RATE OF SEDIMENT TRANSPORT

Having determined the type of sediment that is likely to be moved, or transported, along a given stretch of coastline, it is equally important to estimate the rate at which it is being moved.

The rate of sediment transport is simply the mass of sediment that is moved over a given distance in a set time. This sort of information is needed, for example, when breakwaters are to be built to prevent the loss of beach sands. To calculate the sediment transport rate, we have to consider both the rate at which the bedload is moved (q_b) and the rate at which suspended sediment is moved (q_s).

How would you expect the ratio of q_s to q_b to change as a current speed increased across a bed of mixed grain sizes?

The ratio would increase because as the shear velocity reached values necessary to lift increasingly coarser grains permanently into suspension, more sediment would be transported in suspension, and less as bedload.

4.3.1 THE BEDLOAD TRANSPORT RATE

It has been found both experimentally, and from theoretical considerations, that the rate of bedload transport (q_b) is proportional to the cube of the shear velocity, i.e.:

$$q_b \propto u_*^3 \tag{4.9}$$

provided that τ_0 (the shear stress operating at the bed) is greater than τ_c (the critical shear stress).

The reason for this can be seen from the following argument. When a current expends energy moving sediment along the bed, the current is doing work. The amount of work done depends upon the power available, which is defined as the rate of doing work (in other words, the rate of expending energy). In simple terms, the force exerted by the current at the bed, τ_0, is proportional to the current's energy content. And the power of the current, or the rate at which this energy is being propagated, must be proportional to $\tau_0 \times$ current speed or $\tau_0 \times \bar{u}$. Now, since $\tau_0 \propto u_*^2$ (equation 4.6), power $\propto u_*^2 \times \bar{u}$. Because $\bar{u} \propto u_*$, then power $\propto u_*^3$, $q_b \propto u_*^3$, and $q_b \propto \bar{u}^3$. This relationship is important because it means that very small changes in the current speed or, say, the bed roughness, can have significant effects on the rate of bedload transport. We can illustrate this with a practical example.

Figure 4.10(a) shows the changes in tidal current velocity measured at one metre above the bed during a complete tidal cycle at a location in the North Sea. The south–south-westerly currents flow for slightly longer, and attain slightly higher velocities than the north–north-easterly currents. As a result of the relationship between q_b and current velocity, appreciable differences occur between the amounts of sediment that can be transported in each tidal current direction (Figure 4.10(b)).

QUESTION 4.9(a) What is the difference between the maximum current velocity of the NNE tidal current, and that of the SSW current?

Figure 4.10(a)   The changes in current velocity at 1 m above the sea-bed ($\bar{u}_1$) during a complete tidal cycle at a location in the North Sea. The small gap in the time-scale between +6 and −6 hours is due to the fact that the tidal cycle is about 12.5 hours long. The dashed line is the threshold velocity required to move sand grains of 0.3 mm diameter in the bedload.

(b)   The changes in $\bar{u}_1{}^3$ with time during the same tidal cycle. The shaded areas are proportional to the amount of sediment transported. The horizontal dashed line represents the cube of the threshold velocity shown in (a). LW and HW are low and high water respectively.

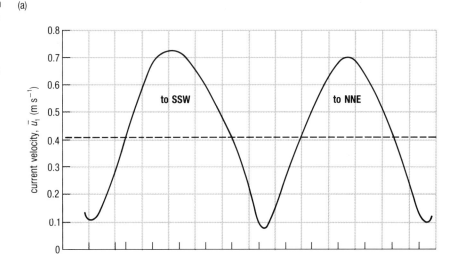

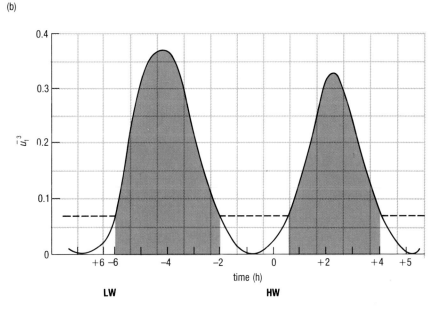

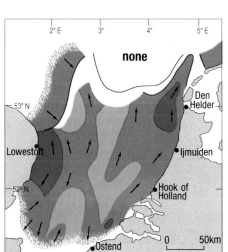

theoretical net sand transport

high
moderate
low

Figure 4.11   Theoretical net transport of sand by tidal currents in the southern North Sea using threshold velocities for grain diameters of 0.2 and 0.3 mm. Arrows show direction of net transport.

(b)   Estimate the difference between the sizes of the shaded areas in Figure 4.10(b), and hence suggest whether there is likely to be significant net transport of sediment in one particular direction at this location.

(c)   Examine the dashed horizontal lines and shaded areas in Figure 4.10(b) (see the Figure caption for an explanation of the lines and areas). What can you conclude about the type of sediment that is being transported at the current velocity represented by these lines?

By repeating the procedure shown in Figure 4.10 for many other locations in the North Sea, a map can be compiled of patterns of theoretical net transport of bedload sediments. Figure 4.11 shows the results of such an exercise, and is a good approximation to the observed sediment movement pattern in the southern North Sea.

It is very difficult to measure physically the actual rate of bedload transport in the sea, because by definition the movement takes place at, or very close to, the interface between the sea-bed and the overlying water. If we try to sample the bedload, the likelihood is that we shall also sample some of the suspended load, and even sediment from the bed beneath the moving bedload. During the mid-1980s, various techniques were developed which all proved too rudimentary to be reliable. Nevertheless, it is still interesting to see what some of these involve.

One method uses the movement of sediment ripples at the sea-bed. These migrate as the sediment moves. So, if we know the size of the ripple, we can calculate the mass of sediment that is moving over a given length of the ripple crest and then work out the rate from how fast the ripple is migrating. To do this, a rod, about one metre long, is held horizontally into the flow about ten centimetres above the bed. A light is shone obliquely onto the rod from one side so that the shadow of the rod on the bed appears as a zig-zag line where it is distorted by the ripple. The progression of the sand ripple is measured photographically by recording the changing size and shape of the shadow. As the method allows for only a few ripples to be measured, the results are unlikely to be representative of the overall pattern of sediment transport in an area.

Another approach has been to detect the self-generated noise produced by grains colliding during sediment movement. The intensity of the sound is related to both the amount of sediment and the grain size that is moving. This method is most suitable for coarser grain sizes but there are problems in calibrating the signal responses from the recording device.

4.3.2   THE RATE OF TRANSPORT OF THE SUSPENDED LOAD

The determination of the rate of suspended load transport is straightforward by comparison with measuring the rate of bedload

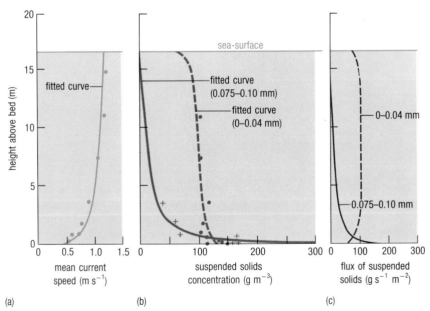

Figure 4.12   Profile of (a) mean current speed, and (b) concentrations of sediment grains with diameters in two different size ranges, multiplied to give (c) the sediment flux throughout the water column. Note that $(ms^{-1}) \times (gm^{-1})$ gives $gs^{-1}m^{-2}$, a flux per unit area; part (c) does not show the total suspended flux but only the fluxes for those particle sizes shown in (b).

transport, but it still has its problems. Current speeds (Figure 4.12(a)) and sediment concentrations (Figure 4.12(b)) are measured throughout the water column and then q_s is calculated by multiplying together the two sets of data. Traditional methods of determining the concentration of suspended sediment by direct sampling of the water column are laborious, expensive and inaccurate. Research is continuing into the development of probes which operate using back-scattered or forward-scattered ultrasound (optical backscatter probes). Another technique uses electromagnetic flow meters in conjunction with fast-response sensors (or transmissometers) which record the impact of sediment grains.

QUESTION 4.10   What can you conclude from Figure 4.12(b) and (c) about the way in which sediment concentrations and the suspended sediment fluxes for grains of different sizes vary through the water column?

4.4   THE DEPOSITION OF SEDIMENT

So far, we have considered only the flow conditions that lead to the movement or erosion of sediment. Water movements in the marine environment are not steady, as you have seen already (Section 4.1.6); tidal currents accelerate to a maximum speed, before decelerating again.

Do you recall why most sediment transport occurs shortly *after* maximum speeds have been achieved?

Remember that as a current accelerates (or decelerates) the shear stress at the bed lags behind that predicted by the current flow observed close to the surface (Section 4.1.6).

After attaining their maximum speeds, tidal currents then decelerate and consequently turbulence at the bed is no longer sufficient to keep all the sediment moving either in suspension or as bedload.

Similarly, waves vary in energy on a seasonal and even a daily basis and so their capacity to move sediment varies. In both cases, a reduction in the capacity to move sediments leads, as you would expect, to the deposition of some of the load being carried. The conditions that govern deposition from the bedload are different from those for the suspended load, and so we shall consider them separately.

4.4.1   DEPOSITION OF THE BEDLOAD

Only coarser sediment (grains larger than about 0.1 to 0.2mm diameter) is transported by movement over the sea-floor as bedload. For a given grain size, these grains will stop moving as soon as the bed shear stress (τ_0) falls below the critical shear stress (τ_c) that was needed to start them moving. In this case, the **critical depositional shear stress**, τ_d, is effectively the same as τ_c.

4.4.2   DEPOSITION OF THE SUSPENDED LOAD

When the critical depositional shear stress is reached for sediment grains in suspension, they will begin to settle towards the bed.

Is the value of τ_d likely to be the same as that of τ_c for suspended grains?

If you look back at Figure 4.9, and consider the discussion in Section 4.2.1, you will appreciate that it is not the same for coarser grain sizes which go through a stage of bedload transport. Grains coarser than about 0.1mm will begin to settle as soon as gravitational forces exceed buoyancy forces but are likely to be intermittently kept in suspension as turbulent eddies fluctuate. If the shear velocity of the current continues to decrease, the grains coarser than about 0.2mm will eventually settle on the bed but continue to move as part of the bedload. So, τ_d at which grain settling begins must be greater than τ_c required to start them moving in the first place as bedload. Grains finer than 0.1mm do not go through a stage of bedload transport and so for such grains τ_d is likely to be closer to τ_c.

As a current decreases, and τ_d for a given grain size in suspension is reached, grains of this size do not all reach the bed at the same time, or even approximately the same time. This is because they will be distributed at different depths in the water column (Figure 4.10(b)). The time they take to settle will depend to a large extent on their settling velocities, and on turbulent eddies, which will have localized, vertically upward velocities sufficient to counteract the settling velocities.

QUESTION 4.11   Refer back to the answer to Question 4.10, and the information about settling velocities in Figure 4.9. Calculate roughly how long it would take for most of the sediment 0.1mm in diameter shown in Figure 4.12(b) to be deposited when the critical depositional shear velocity is reached. (Assume that there are no turbulent eddies impeding the settling of the grains.)

As the settling velocities of sediment grains are directly related to their grain size, very small particles settle significantly more slowly than coarse ones which means they may not reach the bed until well after the current transporting them has dropped below the critical depositional shear velocity—a phenomenon known as **settling lag**. This process is important for sediment deposition on the mud-flats of estuaries and tidal flats. Also, grains of slightly different sizes settle at very different rates because for fine grain sizes $w_s \propto D^2$, and a very small decrease in grain size in the clay to very fine sand range results in a significant change in settling velocity. Therefore, the finer the grain size, the more significant settling lag becomes. Conversely, for grains coarser than about 2mm, $w_s \propto D^{1/2}$, and so even quite large changes in diameter result in only very small variations in settling velocity and there is less separation in grain size of the sediment during deposition. This will become apparent if you compare the grain size and settling velocity scales on Figure 4.9.

4.4.3   RATES OF SEDIMENT DEPOSITION

As water flow conditions in a shallow marine environment change, it is important to know the rate at which sediment is likely to be deposited if flow decreases and the rate at which it will be eroded if flow increases. Some idea of potential deposition rates from the bedload can be gained if the rate of bedload transport is known. As $q_b \propto u_*^3$ (equation 4.9), we may assume that the transport rate will decrease (and so the rate of deposition will increase) in proportion to the reduction in u_*^3 and hence to the reduction in $\bar{u}^3$.

However, from the discussion in the previous Section, it is apparent that the rate at which suspended sediment is deposited (R_d) depends on more than just the decrease in current speed. The rate of deposition will be greater if most of the suspended sediment is concentrated close to the bed, rather than distributed higher up the water column. It will also increase as the settling velocities (w_s) of the sediment grains increase, and as the ratio between the shear stress at the bed and the critical depositional shear stress required to deposit grains of a given size (τ_0/τ_d) decreases. These factors can be combined into a single equation, based on experimental observation, which gives an approximation of the rates of deposition for grains of a given settling velocity:

$$R_d = C_b w_s \left(1 - \frac{\tau_0}{\tau_d}\right) \qquad (4.10)$$

where C_b is the concentration of sediment within a few tens of centimetres of the bed.

It should be apparent from equation 4.10 that the value of $(1-(\tau_0/\tau_d))$ would be negative only if the shear stress at the bed were greater than the critical depositional shear stress. In this case, the value of R_d would also be negative and this would imply that erosion of the grains, and not deposition, was taking place.

If the concentration of sediment (C_b) is measured in $kg\,m^{-3}$, the units of R_d must be $(kg\,m^{-3}) \times (m\,s^{-1})$, i.e. $kg\,m^{-2}s^{-1}$. In other words, values of R_d represent a mass of sediment deposited over a given area of bed within a particular time.

Question 4.12 tests your understanding of the relationships shown in equation 4.10 through some relatively simple calculations.

QUESTION 4.12   A current flowing over a sheltered part of the sea-bed transports mostly fine silt, around 0.01 mm grain diameter. The concentration of this sediment in suspension close to the bed has been measured as $0.005\,kg\,m^{-3}$. A local reduction in the current speed leads to a decrease in the shear velocity to $0.002\,m\,s^{-1}$, and consequent deposition of the sediment.

(a)   Calculate the rate of deposition of the sediment. (You will need to use both Figure 4.9 and equation 4.6; assume that the density of seawater is $1000\,kg\,m^{-3}$.)

(b)   What would be the *annual* rate of deposition of sediment at this place, assuming the local reduction in current speed was maintained?

(c)   What would the rate of deposition depend on if the current speed fell to zero (effectively, still water), and what would the value of R_d be in this case?

As we said, equation 4.10 provides only a rough approximation. In the real marine environment, various other factors may either increase or decrease the value of R_d, such as the degree of bed roughness, the resuspension of sediment by marine organisms, and occasional brief episodes of erosion during an overall period of net deposition.

Figure 4.13 Typical asymmetrical ripples formed by tidal currents in an estuary.

4.5 BED FORMS

The sea-bed is rarely flat (Sections 4.1.5 and 4.1.6), but is more usually covered with both small- and large-scale sediment features. The most familiar of these are probably the small-scale current ripples seen in the muds of estuaries and on tidal flats (Figure 4.13) and the wave-formed ripples common on beaches. These features are known as **bed forms** and they range in size from these small ripples up to the vast sand-waves and sandbanks beneath the shelf seas which may be built up more than 15m above the sea-floor. We shall discuss the formation of ripples in sediment by wave action, and the larger sand-waves and sandbanks, when we consider the beach and shelf environment in Chapters 5 and 8, respectively. In this Chapter, we confine ourselves to those bed forms that develop in shallow coastal waters as the result of currents rather than waves.

4.5.1 THE FORMATION OF CURRENT-PRODUCED BED FORMS

Current-produced bed forms are usually asymmetrical with the steeper slope facing downstream. During active sediment transport, sediment is moved mainly as bedload up the shallower slope, and redeposited down the steeper slope. In this way, both the sediment and the bed form migrate across the sea-bed. You may recall that the rate of migration of ripples has been used to try to determine the rate of transport of the bedload (Section 4.3.1). After transport ceases, bed forms become static and, if buried by further layers of sediment, they may become preserved within the sedimentary record, although not always in a perfect state. It is not unusual to find the bed forms produced by, say, a flood tidal current partially eroded and flattened by the following ebb tide.

4.5.2 CURRENT FLOW AND BED FORMS

Broadly speaking, the types of bed form produced are related to the speed with which a current flows. Small-scale ripples are produced by slower currents than are somewhat larger bed forms. However, the picture is more complicated than this, because both the grain size of the sediment and the depth of water influence the size and shape of bed forms. Figure 4.14 shows the relationship between current-produced bed form, mean flow velocity and sediment grain size for a flow depth of about 0.4m.

Current ripples develop at relatively slow current speeds and occur only in sediments finer than about 0.6mm grain size. The water flow is only slightly disturbed above the ripples (Figure 4.15(a)). It seems that they can only form when the grain roughness does not totally destroy the viscous sublayer. Since the thickness of the sublayer decreases as the speed of a current increases, it follows that ripples are unlikely to develop at fast current speeds or when there is coarse sediment on the bed.

At higher current speeds, and in coarser sediments, somewhat larger bed forms known as megaripples* are produced. These are up to a metre or more in height, and have wavelengths often of several metres or even tens of metres (Figure 4.16).

*You will find the term 'dune form' used in some textbooks written primarily for geologists. However, marine sedimentologists prefer the name 'megaripple' as this avoids confusion with features such as wind-blown sand-dunes.

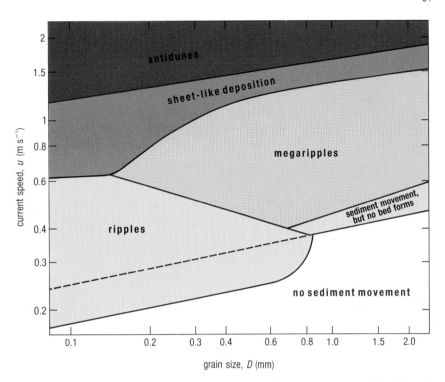

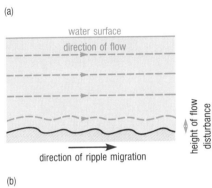

(a)

water surface

direction of flow

height of flow disturbance

direction of ripple migration

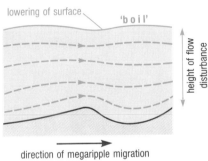

(b)

lowering of surface

'boil'

height of flow disturbance

direction of megaripple migration

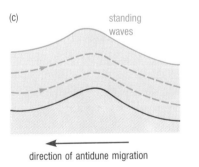

(c)

standing waves

direction of antidune migration

Figure 4.14 The relationships between bed forms, mean current speed, and sediment grain size based on experimental data for a water flow depth of about 0.4m. In the region below the dashed line, ripples will not form spontaneously from a flat bed—the bed must be uneven in the first place. However, once formed, these ripples will continue to migrate.

Figure 4.15 The flow of water (dashed lines) over ripples, megaripples and antidunes, and the direction of migration of the bed forms (solid arrows).

(a) Ripples: there is only minor disturbance of the flow by the bed forms and this occurs directly above the ripples.

(b) Megaripples: the flow is disturbed up to the flow surface where 'boils' can be seen developing.

(c) Antidunes: a standing wave develops at the surface of the flow, in phase with the antidune.

The water flow above a megaripple is usually disturbed all the way up to the surface (Figure 4.15(b)) and the slightly raised area at the surface is usually recognized as turbulence, or 'boils', where water moves upwards and spills outwards. You will notice from Figure 4.14 that when the current is sufficiently strong, it wipes out bed forms altogether and sediment is moved in sheet-like layers over a plane bed. At even higher current speeds, antidunes develop (Figure 4.15(c)). Erosion of antidunes occurs on the shallow slope, and deposition on the steep slope, so that the *shape* of the bed form actually migrates *upstream*, although the sediment itself moves downstream. As with megaripples, when antidunes develop, the water flow is visibly disturbed all the way to the surface where standing waves develop; these are in phase with the bed form and do not progress across the water surface like wind-generated waves, but move only as the bed form migrates (Figure 4.16(c)). Whilst ripples and megaripples are very common coastal and shallow marine bed forms, antidunes are rarely permanent in the marine environment because of the high current speeds needed to form them. However, you may be able to observe them forming where swell-wave water or a fast-flowing stream crosses a beach, and the water flow is sufficiently shallow for standing waves to develop at the surface (Figure 4.16(c)).

The shapes of bed form crests are also related to flow conditions. Where flows are relatively slow, or deep, the bed forms are linear with long straight crests, like the ripples shown in Figure 4.13. However, at high speeds, or in shallower water, the crests become progressively more indented until eventually they are broken up into short, curved sections, like the ripples shown in Figure 4.16(a).

(a)

(b)

(c)

Figure 4.16(a) Current-formed ripples with curved crests.

(b) Megaripples with superimposed, curved crested current ripples.

(c) Antidune formation (centre of picture) in beach sands crossed by a fast-flowing stream. Wave crests are ~ 50cm apart.

4.6 SUMMARY OF CHAPTER 4

1   Movement of sediment occurs when the force of water flowing over the sediment is able to overcome the force of gravity acting on the sediment grains and the friction between the grains and the underlying bed.

2   Friction between the bottom layers of water and the bed produces a shear stress at the bed. Frictional retardation with the bed generates a boundary layer. The shear stress is directly proportional to both the velocity gradient within the boundary layer and the viscosity of the water.

3   Natural water flows, including flows in the oceans, are turbulent rather than laminar. In turbulent flow, bed shear stress is proportional to the square of the mean flow speed. When water flows over a smooth (very fine-grained) bed, flow is said to be smooth turbulent. The basal layer appears to flow in a laminar fashion forming a viscous sublayer only a few millimetres thick which decreases in thickness as flow speed increases. When water flows over a coarse-grained bed, or at high current speeds, grains are able to protrude through the viscous sublayer, breaking it

down, and flow is said to be rough turbulent. Turbulent eddies are able to reach down between sediment grains giving rise to greater potential for sediment movement.

4   Sediment movement at the bed is often considered as a function of the shear velocity rather than of the shear stress. The shear velocity is a term derived from the square root of the ratio between the shear stress and the density of the flow, but may be calculated directly from the velocity gradient, $d\bar{u}/dz$. The intercept of the logarithmic velocity gradient $d\bar{u}/dz$ on the depth axis gives a measure of the bed roughness length (z_0) which increases as the sediment grain size increases; it will also be greater if there are features such as ripple marks present.

5   In the marine environment, values of bed shear stress may be overestimated or underestimated as the result of decelerating and accelerating tidal currents respectively; the time taken for a velocity profile to adjust to a new bed roughness; and the concentration of suspended sediment close to the bed. Not all the shear stress estimated from a velocity profile is available to move sediment. Resistance due to form drag reduces the capacity for sediment movement.

6   Movement of grains of a given size begins when the shear stress at the bed (and consequently the shear velocity) reaches a critical value. Non-cohesive quartz grains coarser than about 0.1mm in diameter are moved as bedload until the ratio of the grain settling velocity to the shear velocity is about 1.2. They are then lifted intermittently into suspension. At ratios less than about 0.1 they are lifted permanently into suspension. Grains finer than about 0.1mm are lifted into suspension as soon as the critical shear velocity is reached. Erosion of cohesive sediments (muds and clays) that have become partially consolidated occurs by mass failure. The resistance of muds to erosion is assessed by their yield strength.

7   The rate of bedload transport is proportional to the cube of the shear velocity, but is difficult to measure in the sea. At a given height in the water column, the rate of suspended load transport is calculated as the product of the current velocity and the sediment concentration.

8   Deposition of the bedload, or cessation of transport, occurs when the shear stress reaches the critical depositional shear stress (τ_d), which is effectively the same as the critical shear stress required to start the sediment moving (τ_c). Deposition of the suspended load varies according to the sediment grain size. For grains coarser than about 0.1mm, τ_d is greater than τ_c and grains enter a phase of bedload transport until τ_c is reached, when transport ceases. Grains finer than about 0.1mm do not enter a phase of bedload transport and so τ_d is close to τ_c. The settling lag of fine suspended grains prevents them reaching the bed until well after the shear velocity has fallen below the critical depositional shear velocity. The rate of deposition of the bedload is proportional to the reduction in $\bar{u}^3$. However, the rate of deposition of grains of a given size in the suspended sediment load is proportional to the concentration of sediment close to the bed, the settling velocity of the grains, and the ratio between the actual shear stress at the bed and the depositional shear stress.

9   Raised sediment features on a sediment bed are known as bed forms. Most current bed forms are asymmetrical and migrate downstream in the direction of sediment transport. The type of bed form depends on the

94

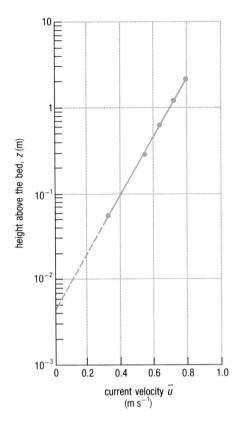

Figure 4.17   A log velocity profile (for use with Question 4.13).

current speed, the sediment grain size and the flow depth. Small-scale current ripples form at relatively slow current speeds and in sediment finer than about 0.6mm grain diameter. Successively higher speeds or coarser-grained sediments produce larger-scale megaripples, then planar beds, and, if the speed reaches a sufficiently high value, antidunes may be formed which migrate upstream in the opposite direction to the sediment transport.

Now try the following questions to consolidate your understanding of this Chapter.

QUESTION 4.13   Figure 4.17 is a log velocity profile for the lower part of the boundary layer, formed where a tidal current flows over the sea-bed.

(a)   Calculate the shear velocity of the tidal current and the shear stress on the bed beneath the flow. (Assume the density of seawater is $1000 kgm^{-3}$.) How does the value of u_* compare with the actual current speed one metre above the bed?

(b)   Why may the values for u_* and τ_0 that you have calculated in (a) be unreliable?

(c)   Assuming the values *are* reliable, predict the approximate range of grain sizes which could be moved only as bedload. Which types of sediment does this range include?

(d)   What is the maximum grain size that could be transported fully as suspended load? Which type of sediment does this represent?

(e)   How would the presence of bed forms in the sediments at the bed beneath the tidal flow affect your answers to (c) and (d) above?

QUESTION 4.14   How might the inspection of bed forms on the bed of the southern North Sea help to confirm the net transport paths for sand shown in Figure 4.11?

QUESTION 4.15   Figure 4.16(b) shows a series of large, current-formed megaripples with superimposed, small-scale, current-formed ripples.

(a)   In which direction (relative to the photograph) was the current flowing when the bed forms were developed?

(b)   Explain how small-scale ripples have been formed superimposed on megaripples.

(c)   What can you say about the grain size of the sediment in which these bed forms have developed?

CHAPTER 5   BEACHES AND THE LITTORAL ZONE

'…I gain the cove with pushing prow,
And quench its speed in the slushy sand.
Then a mile of warm sea-scented beach;'
From *Meeting at Night* by Robert Browning.

For most people, the word 'beach' means simply that stretch of sand or shingle above low tide level where they can lay out their sun beds, erect the deckchairs and keep a wary eye on the rising tide while the children paddle or build sandcastles. In other words, their interest lies only in those mainly sandy areas of the shoreline which are exposed at some stage in the tidal cycle. Coastal oceanographers have a wider interest because processes involving water and sediment movement occur below low tide level, and these must be considered if the beach zone is to be understood. In this Chapter, therefore, we shall examine the much broader **littoral zone** which stretches between the seawards limit of land plants and the region below sea-level where sediment is not disturbed by wave action during fair weather conditions—i.e. around 10m to 20m water depth at low tide*. Where cliffs are developed, they (rather than vegetation) limit the landwards extent of the coastal zone. Seawards of the littoral zone is the **offshore zone**.

As beaches are accumulations of loose sand or pebbles, they change shape rapidly in response to changes in wave energy, and the movement of beach sediment dissipates some of the energy of a wave breaking on the shore (Section 1.4.6).

5.1 THE DIVISIONS OF THE LITTORAL ZONE

Rather a large amount of terminology exists to describe the features of the littoral zone. The terminology used depends on whether oceanographers are concerned with the influence on the littoral zone of tidal currents, the effects of waves, or the sediment profile. Figure 5.1 draws all the terms together and shows how they are related.

The part of the littoral zone that is exposed at low water when the tide is out, but covered at high water when the tide is in, is known as the **foreshore** or **intertidal zone**. The **backshore** is above mean high tide and is only influenced by the sea when there are storm waves, or during exceptionally high tides. Sediment on the backshore dries out rapidly and, where the coastline is flat, it is easily worked by the wind into a series of coastal sand dunes. Seawards of the foreshore is the **shoreface** which is permanently covered by water, except at exceptionally low tides.

*Note that marine ecologists use 'littoral' in a different sense, to mean only the intertidal zone.

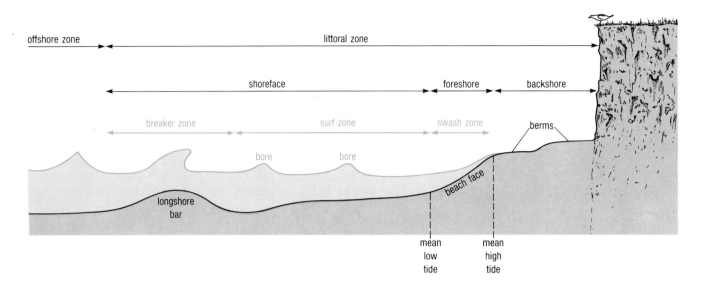

offshore zone — littoral zone

shoreface — foreshore — backshore

breaker zone — surf zone — swash zone — berms

bore — bore — beach face

longshore bar

mean low tide — mean high tide

Figure 5.1   Zones of tidal influence, wave action and the sediment profile within the littoral zone.

5.1.1   ZONES OF WAVE ACTION

On the basis of wave action, the littoral zone is divided into breaker, surf and swash zones. The **breaker zone** is where waves become unstable and break, generating a **surf zone** in which much shallower waves (or bores) are projected up the beach face to the **swash zone**. The swash zone is alternately covered by the upsurge of water (the swash) and exposed as the backwash retreats.

QUESTION 5.1   Are the wave zones constant in position along a stretch of coastline, during a tidal cycle?

5.1.2   THE SEDIMENT PROFILE

Figure 5.1 shows that the sediment in the littoral zone does not form a smooth landwards slope, but is interrupted by a number of ridges or sediment bars. The **berm** of the backshore is a flat-topped ridge which develops at the limit of the wave swash on steeply sloping beaches (Figure 5.2(a)). By contrast, shallow sloping beaches of the foreshore are often characterized by a series of low broad sandy bars separated by linear depressions or **runnels** (Figure 5.2(b)) running parallel to the shoreline. These bars, sometimes known as **swash bars**, form by sediment movement in the surf and swash zones and so a whole series of swash bars and runnels may develop as the tide migrates across the beach. As Figure 5.1 shows a steep beach profile, no swash bars or runnels are drawn. It may be quite dangerous to walk down to the sea's edge on the incoming tide across a beach with well-defined swash bars and runnels, unless you are a strong swimmer. The runnels fill with water first and it is easy to find yourself stranded on a swash bar, separated from the next bar or the shore by a stretch of often relatively deep, fast-flowing water.

Seawards, beneath the breaker zone, a **longshore bar** may develop. It is a characteristic feature of some beach profiles in the winter season, when

(a)

Figure 5.2 (a) A beach berm. (b) A swash bar
and runnel exposed on the foreshore.

(b)

berms are generally absent. We shall discuss the reasons for the formation
of longshore bars when beach profiles are examined in more detail in
Section 5.3.

The sloping portion of the beach, below the berm shown in Figure 5.1, is
the **beach face**.

QUESTION 5.2   Examine the extent of the beach face on Figure 5.1 and
describe (a) the relationship between the beach face and the intertidal
zone, (b) the relationship between the beach face and zones of wave
action.

5.2 SEDIMENT MOVEMENT IN THE LITTORAL ZONE

Sediment movement in the littoral zone is more complicated than that described by the relatively simple model for unidirectional water currents derived in Chapter 4. As a wave propagates through water, particles of water at the surface follow an almost circular orbit. In shallow water, where water depth is less than half the wavelength, the orbits become progressively flattened at the bed (Figure 1.8(c)). This means that near the sea-bed there is movement to and fro of the water. In addition to wave action, there is water movement by tidal currents and, furthermore, wave action itself usually generates water currents which run parallel to the shoreline. To begin with, we shall consider only the effects of wave action.

5.2.1 ORBITAL VELOCITIES OF WAVES AND BED SHEAR STRESS

In deep water, the water particle orbits under waves are almost circular, with the diameter of the orbits decreasing with depth (Figure 1.8(a)). The speed at which the water (as opposed to the wave-form) moves is calculated from the length of time it takes a water particle to complete an orbit; i.e. for the particle to move from crest to trough, and back to the crest of the next wave as the wave-form passes. This speed is known as the **orbital velocity**. The horizontal velocity of the particle is at a maximum (u_m) in a landwards direction just beneath the crest of a wave, and in a seawards direction just beneath the trough. The value of u_m is given by:

$$u_m = \frac{\pi D_0}{T} \tag{5.1}$$

where D_0 is the orbital diameter and T is the wave period.

As a wave begins to enter shallower water, depths less than $L/2$ but more than $L/20$, the water particle orbits begin to flatten and become narrower with depth until at the bed there is simply to and fro movement. The maximum horizontal orbital velocity near the bed occurs below the wave crest and trough again, and is related to the wave height H, water depth d, wavelength L, and wave period T, by:

$$u_m = \frac{\pi H}{T \sinh (2\pi d/L)} \tag{5.2}$$

Sinh is the hyperbolic sine, a mathematical function analogous to the hyperbolic tangent introduced in equation 1.3. The important points to notice in equation 5.2 are the relationships between u_m and T and between u_m and H. For waves of a constant period, as wave height increases so u_m increases. For waves of a given height, u_m is greater for short-period waves than long-period waves.

In water of depths less than $L/20$, waves behave as true shallow water waves and the orbital velocity is constant with depth:

$$u_m = \frac{Hc}{2d} = \frac{H}{2}\sqrt{\frac{g}{d}} \tag{5.3}$$

where H is wave height, c is wave speed, d is water depth and g is the acceleration due to gravity. Note that the second form of equation 5.3 is obtained by substituting $\sqrt{gd}$ for c (equation 1.5) and has the advantage that only two variables, wave height and water depth, need to be measured in order to calculate u_m.

QUESTION 5.3   As waves move into shallow water, what happens to:

(a)   their period?

(b)   their speed?

(c)   their height?

(d)   their maximum orbital velocity?

As waves move into progressively shallower water, not only does the value of H increase, but the value of d decreases and both changes lead to an increase in the maximum orbital velocity of the wave and, consequently, to an increase in the shear stress at the sea-bed, and the potential for sediment movement.

Although there is a wave boundary layer, analogous to the current boundary layer described in Section 4.1, it is very much more complex than for unidirectional flow and so we shall not discuss it in detail. The complexity arises because u_m, and hence the shear stress, reverses direction as a wave passes and so the boundary layer never becomes fully established. The wave boundary layer is also very thin, no more than from a few millimetres to one or two centimetres, compared with between 1m and 10m for the boundary layer produced by current movement in the sea (Section 4.1.1). Where medium-grained sands and finer sediment (i.e. sediments less than about 0.5mm diameter) occur on the sea-bed, the flow is analogous to smooth turbulent flow, but when the sediment is coarser-grained, then the flow is rough turbulent. This is important because, as you will see in the next Section, it affects the threshold orbital velocity at which sediment on the sea-bed will begin to move as the result of wave action.

5.2.2   ONSHORE AND OFFSHORE MOVEMENT OF SEDIMENT BY WAVES

As a wave moves into shallow water, the maximum orbital velocity increases until it exceeds a **threshold velocity** (u_t), at which the sediment on the sea-bed begins to move. As the maximum orbital velocity is related to the orbital diameter and wave period (equation 5.2), it is possible to relate u_t to wave period and the size of particle that can be moved (Figure 5.3). Figure 5.3 is constructed specifically for quartz grains with a density of $2.65 \times 10^3 \text{kgm}^{-3}$. The curves would differ if the density of the grains varied. The break in the 1s curve and the 5s curve is due to the change from smooth turbulent flow to rough turbulent flow mentioned in Section 5.2.1. You will notice from Figure 5.3 that the threshold orbital velocity required to move a grain of a given size *increases* as the wave period increases. For example, u_m must be about 0.25ms^{-1} in order to move a quartz grain 1mm in diameter beneath a wave of 1s period. However, a value of about 0.4ms^{-1} is required to move the same grain beneath a wave of 15s period. The reason for this is the rapidity with which a water particle accelerates to its maximum horizontal velocity. The acceleration is much greater for a shorter-period wave than a longer-period wave, and this leads to greater friction at the bed. Equation 5.2 also shows that there are many combinations of wave period, water depth and wave heights which could produce the critical threshold velocity

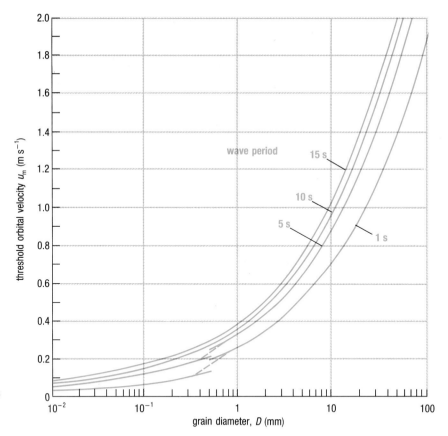

Figure 5.3 The relationship between near-bed maximum orbital velocity and sediment movement under waves of different periods. The breaks in the 1s and 5s curves are the result of the wave boundary layer behaving in a laminar fashion for grain sizes less than about 0.5mm diameter, and in a turbulent fashion for larger grain sizes.

necessary to move a grain of a given size. Figure 5.4 shows the combinations of wave heights and water depths necessary for a wave with period 15s to move grains of the sizes represented, on a flat sea-bed.

QUESTION 5.4(a) If the maximum orbital velocity beneath a wave of a certain height in a given water depth, and with a period of 15s, is about $0.25\,\text{m s}^{-1}$, what is the maximum size of quartz grain that can be moved by the wave?

(b) If the wave height decreased, but water depth remained the same (e.g. after a storm), would you expect these grains to remain in transport?

(c) If the wave height remained the same, but water depth increased (e.g. with the incoming tide), would you expect the grains to remain in transport?

An important point to notice from Figure 5.4 is that large storm waves are capable of moving sediment at considerable depths on the continental shelf. This is important when we consider shelf processes in Chapter 8.

The maximum horizontal velocity is attained twice as a wave passes: once on the forwards stroke of the wave as the crest passes, and secondly on the backwards stroke as the trough passes. This means that sediment should be moved landwards beneath the crest and seawards again beneath the trough.

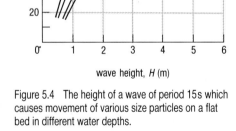

Figure 5.4 The height of a wave of period 15s which causes movement of various size particles on a flat bed in different water depths.

If this is so, does that mean that sediment is moved to and fro on the same part of the sea-bed all the time?

Intuitively, you probably answered 'no' to this question, but you may well not know why there is actually a net movement of sediment either landwards or seawards. The net movement is related to the movement of water particles.

On the backwards stroke of the wave, when a water particle is in the trough, it is brought closer to the sea-bed than when it is at the crest of the wave, therefore frictional retardation due to the bed is greater. This means that water particle speeds are not the same in both directions. During onshore movement they are at their highest, but this speed is maintained for only a short period of time. During offshore movement, the speeds are slightly lower, but are maintained for longer periods of time (Figure 5.5). During onshore movement, coarser sediment is moved as bedload and finer sediment as suspended load.

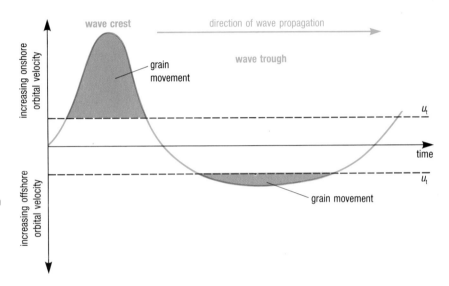

Figure 5.5 The asymmetry of water particle velocities associated with a shallow-water wave (curved line). u_t is the threshold velocity at which grains of a given size will be set in motion. The brown shaded area under (or above) each curve represents the range of velocities over which grains of that size will be transported. The unshaded areas represent the range of velocities over which these grains will not be transported.

QUESTION 5.5   How is this inequality of flow likely to affect the seawards distribution of sediment grain sizes in the littoral zone?

During offshore movement, not all the coarse material can be returned seawards and so only the finer bedload and suspended load are returned. As the lower offshore speeds are maintained for longer periods of time, there will be a net offshore movement of fine material, and a net onshore movement of coarse material.

5.2.3   THE LONGSHORE TRANSPORT OF SEDIMENT BY WAVE-GENERATED CURRENTS

Two wave-induced current systems may be recognized, both of which generate currents parallel to the shoreline, or **longshore currents**, and therefore move sediment along the shoreline. The first type is produced by wave crests approaching the shore obliquely instead of parallel to the shoreline, and the second is associated with the development of rip currents. When, as often happens, both current systems occur together, they interact.

Longshore currents due to oblique waves
When waves approach a straight coastline at an oblique angle (such an angle being the most common situation), a longshore current is

established which flows parallel to the shoreline in the nearshore region with speeds between about 0.3 and $1\,\mathrm{ms^{-1}}$. The speeds of these currents are proportional to both the maximum orbital velocities of waves in the breaker zone, and the angle that the wave-fronts make with the shoreline as they approach it. These longshore currents are best developed along straight coastlines and are an important way of moving sediment along shorelines where there are gently sloping beaches.

On steep-faced beaches, transport by swash and backwash is more important than by longshore currents. When a wave breaks obliquely to the shoreline, the swash drives sediment up the beach face at an angle to the shoreline. However, the backwash drags the sediment down the beach face at right angles to the shoreline and successive waves move the sediment along in a zig-zag pattern.

Rip currents
Rip currents are strong, narrow currents with speeds up to $2\,\mathrm{ms^{-1}}$ which flow seawards from the surf zone (Figure 5.6). They are potentially very dangerous because a swimmer caught in a rip current may be swept out to sea quite rapidly and drown after becoming exhausted by trying to swim back to the shore against the current. The best means of escape is to swim parallel to the shore for a few metres, away from the rip current, before trying to swim shorewards. However, experienced surfers are quite happy to exploit these currents by riding them out to sea.

The rip current forms part of the cell-like circulation of water shown by the arrows in Figure 5.6(a). Longshore currents move towards each other in the surf zone, and where they converge, the water turns seawards as a rip current. The longshore currents that drive the rip are believed to result from variations in wave height along a wave crest, or series of wave crests, breaking along a beach. The presence of waves in the nearshore zone causes the water level to deviate from the horizontal surface it would have if the water was completely still (Figure 5.7). Just shorewards of the breaking waves, the average water level falls and then rises continuously towards the shore, an increase in level known as **wave set-up**. Where wave heights are greatest along the wave crest, the wave set-up is also greatest and so horizontal pressure gradients exist between regions of high wave set-up and those of lower wave set-up. Consequently, water moves from positions of highest wave height to those of lowest wave height, generating the longshore currents which feed the rip currents (Figure 5.6(a)).

As you know, an increase in wave height along a wave crest occurs when the wave enters shallow water and begins to slow down. Variations in wave height along a wave crest will occur if one section of the crest encounters shallow water before another. This will occur either when a wave crest approaches an irregular coastline and shallow water is encountered first off seawards projections, or where the nearshore submarine topography is irregular for other reasons. Once initiated, the rip current itself, moving seawards in the opposite direction to the wave crest, will cause a local increase in height of the incoming waves (Section 1.5.1).

During rip current circulation, sediment is moved along the shore by the longshore currents, and seawards by the rip current (Figure 5.6(b)). It has been suggested that rip currents may be the link between the circulation

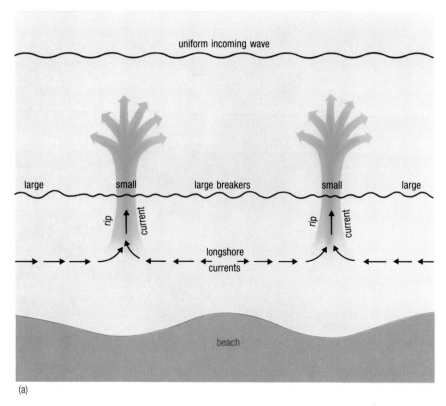

(a)

Figure 5.6(a) Plan view of a section of coastline showing the formation of rip currents due to the variation in wave heights along wave crests.

(b) In this aerial photograph, the presence of rip currents seawards of the breaker zone is indicated by the sediment transported offshore (light-coloured patches).

(b)

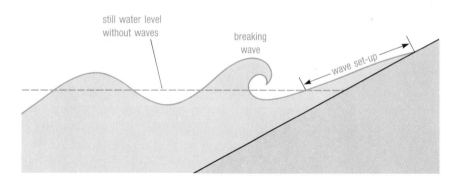

Figure 5.7 Wave set-up: the rise in the mean water level inshore of a breaker.

of sand on the shoreface and its deposition beyond the shoreface in the offshore shelf region. Rip current circulation may also be an important means of renewing water in the nearshore zone and flushing out sewage and other pollutants dumped in coastal regions.

5.2.4   THE LONGSHORE SEDIMENT TRANSPORT RATE

Coastal engineers designing jetties and breakwaters need to know the rate at which sediment is being moved along the shoreline so that they can estimate how these structures may modify the sediment transport. Before the longshore transport rate can be estimated, the wave power (P_1) available for longshore transport must be evaluated. Wave power depends on the height of the waves at their break point (H), their group speed (c_g) and the angle between the advancing wave crest and the shoreline at the break point (α) (Figure 5.8).

$$P_1 = c_g(\tfrac{1}{8}\rho gH^2) \sin \alpha \cos \alpha \qquad (5.4)$$

where ρ is the density of the seawater and g is the acceleration due to gravity.

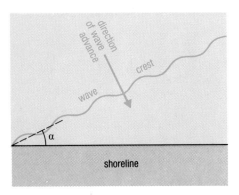

Figure 5.8   The relationship between an oblique wave crest (a), approaching a shoreline (c) at an angle α.

You should recognize the expression in brackets as defining wave energy (E; equation 1.10) and c_g as the speed at which the energy is propagated. The sine term in equation 5.4 is the one that determines the longshore component of wave power per metre crest length of wave. To be meaningful, however, this needs to be converted into power per metre length of shoreline, which is the reason for using the cosine term (*cf.* Figure 5.8).

From equation 5.4 we can derive the base units for wave power as follows: $P_1 = (ms^{-1}) \times (kg\,m^{-3}) \times (m\,s^{-2}) \times (m^2) = kg\,m\,s^{-3}$ which is the same as saying $kg\,m^2s^{-2}s^{-1}m^{-1}$, in other words joules per second per metre, or watts per metre.

If you were to stand on a beach, watching waves breaking onto the shore, you could use equation 5.4 to arrive at a first estimate of wave power in the beach zone by making a few simple observations and estimates of basic wave characteristics. However, you would have to remember that group speed is the same as wave speed in shallow water, and to use the significant wave height (Section 1.1.4) estimated at the break point of the waves.

QUESTION 5.6   Given that wave group speed $= 0.5\,ms^{-1}$, significant wave height $= 1\,m$, and using $\rho \approx 1000\,kg\,m^{-3}$ and $g = 9.8\,ms^{-1}$, calculate (a) the longshore wave power when the wave crest approaches the shoreline at an angle of 30°, (b) the longshore wave power when the wave crest approaches parallel to the shoreline.

The wave power for waves breaking directly on the beach is, therefore, simply:

$$P = c_gE, \text{ i.e. } P = c_g(1/8\,\rho gH^2) \qquad (5.5)$$

Having calculated the wave power, the longshore rate of sediment movement for sand-sized grains, q_1, can be determined from the empirically derived equation:

$$q_1 = \frac{0.77P_1}{g(\rho_s - \rho)\,0.6} \qquad (5.6)$$

where q_l is measured in m^{3}s^{-1}, ρ_s and ρ are the densities of sediment and water respectively, 0.77 is a coefficient of efficiency relating to loss of water due to percolation through the sediments, and 0.6 is another coefficient which represents the average proportion of the bulk sediment occupied by particles, rather than pore space. The longshore transport of sediment is often quite substantial. For example, along parts of the south-eastern coast of the USA, the net southwards transport rate past a point may be as much as 0.5×10^6m^3 per year.

If estimates could be made of the amount of sediment moved by different means in the littoral zone, a sediment budget could be drawn up of material imported (i.e. credited) and material exported (i.e. debited— Figure 5.9) and the balance used to determine whether the beach at any one place is undergoing active deposition or erosion. In practice, this is an almost impossible task given the uncertainties and assumptions in making calculations based on equations similar to 5.5 and 5.6. However, the *likelihood* of beach erosion or deposition can be estimated. An important application of the sediment budget involves the likely effects of any attempt by industry to alter the dynamic equilibrium of the littoral zone.

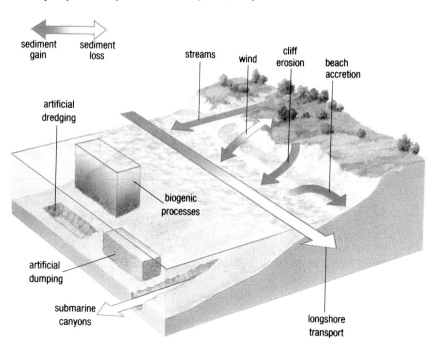

Figure 5.9   The sediment budget in the littoral zone.

5.2.5   SEDIMENT TRANSPORT IN COMBINED WAVES AND CURRENTS

Although we have confined our discussion in preceding Sections to the influence of wave action, in practice waves and currents act together in the littoral zone. It seems likely that most net sediment transport will occur when movement by currents is enhanced by wave motion. Waves are very effective at stirring up sediment on the sea-bed because the orbital motions of the water particles act rather like additional turbulent eddies. Waves can lift sediment of a given grain size into suspension at much lower equivalent speeds than a steady current. Once waves have lifted sediment into suspension, it can then be transported by currents which would, by themselves, be unable to lift the sediment off the sea-bed, before it settles back. These currents include the longshore currents

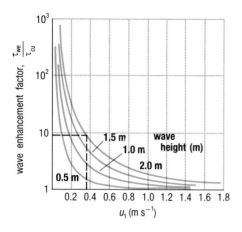

Figure 5.10 The enhancement of currents by waves of various heights, with an 8 s period in water 20 m deep. τ_{cu} is the shear stress under currents alone and τ_{wc} is the shear stress beneath both waves and currents. τ_{wc}/τ_{cu} is the enhancement factor. u_1 is the current speed measured at 1 m above the bed.

discussed in Section 5.2.3, the slight landwards movement of water caused by the wave drift described in Section 1.2.1, and also tidal currents.

The relative importance of sediment movement by waves increases as the influence of current action decreases. Figure 5.10 shows how the wave enhancement varies with the current speed, measured at 1 m above the sea-bed. τ_{cu} is the shear stress which would result from the current alone and τ_{wc} is the shear stress due to both waves and currents. The slower the current, the less τ_{cu} will be and so the larger the enhancement factor, τ_{wc}/τ_{cu}, will be. You should note that Figure 5.10 refers specifically to waves with a period of 8 s in water of depth 20 m.

QUESTION 5.7(a)   Examine Figure 5.10 and consider a wave with height 2 m. How does the enhancement factor change when the current speed at 1 m above the bed is about $0.35\,\mathrm{m\,s^{-1}}$, and when it is about half this speed?

(b)   When the current speed at 1 m above the bed is about $0.35\,\mathrm{m\,s^{-1}}$, what is the difference in the enhancement factor between a wave with height 2 m, and a wave with height 1 m?

5.3 BEACH PROFILES

If you visit the coast frequently, you will almost certainly have noticed that beaches made of coarse sand, shingle or pebbles are steeper than those made of fine sand. You may also be aware that in the summer, when fair weather conditions normally prevail, the beach slope may be steeper than in winter when the sea is often stormy. Clearly both sediment grain size and wave type affect the beach profile.

5.3.1 BEACH PROFILE AND GRAIN SIZE

When a wave breaks on the shore, sediment is pushed up the beach face by the swash, and dragged back down by the backwash. Due to percolation of water into the beach face, the backwash tends to be weaker than the swash. Consequently, there is a net onshore movement of sediment up the beach face until eventually the slope reaches a state of dynamic equilibrium and as much sediment is moved landwards as is returned seawards. The rate of percolation is controlled mainly by the grain size. Water percolates much more easily into a shingle or pebble beach than into a fine sandy beach, and so the backwash is greatly reduced in strength and thus the beach slope is much greater.

5.3.2 BEACH PROFILE AND WAVE TYPE

You have already seen that different types of breakers are associated with the angle of beach slope (Section 1.4.6 and Figure 1.16). Small gentle waves and swell tend to build up beaches whereas storm waves tend to tear them down and flatten them. The controlling factor appears to be the wave steepness.

QUESTION 5.8(a)   Can you recall how wave steepness is defined?

(b)   How does wave steepness vary among the four types of breakers discussed in Chapter 1: spilling breakers; plunging breakers; collapsing breakers; surging breakers?

When a steep wave breaks onto the beach, its energy is dissipated over a relatively narrow area and the swash does not move far up the beach. This means there is less opportunity for percolation to occur and less of the energy is lost in moving sediment up the beach face. Consequently, the backwash is strong and much material can be moved seawards. When a less steep wave, such as a collapsing or surging breaker, breaks on the beach, there is a great deal of movement of water up the beach face as the front of the wave either collapses or surges up the beach. In this case, the swash is strong and so a good deal of sediment is moved up the beach face, too. There is more opportunity for water loss by percolation, and much energy is lost in moving sediment, so the backwash is weak.

Where there is a seasonal difference between fair weather, swell-dominated waves in summer, and steep storm waves in winter, sediment is moved up the beach face during the summer to build berms, while during the winter the berms may be destroyed, and the beach slope flattened. The sediment that is removed from the beach face during winter is transported seawards to build the longshore bar. During the following summer, when swell waves prevail again, the sediment of the bar is moved landwards, up the beach face to reconstruct the berms. Clearly, on any particular beach both the predominant grain size and the prevailing wave steepness will play a part in determining the beach slope. The relationship between these three variables is shown in Figure 5.11.

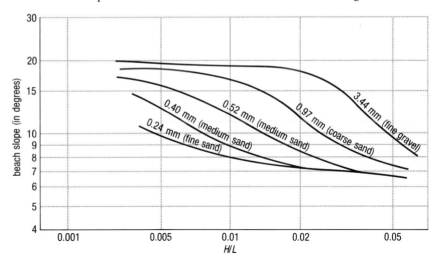

Figure 5.11 The relationship between beach slope (measured in degrees), wave steepness (H/L) and average grain size. Note that the scales for both beach slope and wave steepness are logarithmic.

QUESTION 5.9 The curves in Figure 5.11 show the relationships between beach slope and wave steepness for sediments of different grain sizes. Describe how the influence of wave steepness on beach slope varies as the sediment grain size decreases.

5.4 BEACH MATERIALS AND SEDIMENTARY STRUCTURES

Before leaving this discussion of the littoral zone, it is appropriate to say something about the different types of sediments found there, as well as saying something about the more common depositional features seen in the intertidal zone.

5.4.1 BEACH MATERIALS

Beaches bordering continental areas in temperate climatic zones tend to be formed of pale yellow to brown quartz-rich sands, the most common solid products of weathering and erosion. Pebble or shingle beaches also occur in which the pebbles are usually derived from some fairly local source such as an adjacent line of cliffs. However, quartz beach sands and pebbles are not the only types of sediment found on beaches.

Around volcanic islands, beaches often consist entirely of basaltic or andesitic lavas and may be black in appearance. The green sands round parts of Hawaii contain a high percentage of olivine crystals derived from the surrounding volcanic rocks. In tropical regions, where biological productivity may be high and land-derived sediment scarce, beaches are often comprised of brilliant white sand and gravel-sized fragments of coral and shells, and even carbonate grains which have been precipitated inorganically from seawater. Very occasionally, some beach sediments are artificially derived. For example, beaches near coal-mining districts often contain a high proportion of sand-sized coal fragments, and the pebble beach at Lynmouth in Devon contains pieces of wave-rounded bricks and tiles, debris resulting from a flood disaster in 1952 during which entire houses were washed down the Lyn valley into the sea.

5.4.2 SEDIMENTARY STRUCTURES

One of the most common features seen on the beach is the wave-generated ripple. These ripples are symmetrical and have long, straight crests which occasionally bifurcate. These features distinguish them from the asymmetrical ripples produced by unidirectional currents. They are symmetrical because sediment is moved towards the crest on both sides of the ripple: on the seawards-facing side as the crest of a wave passes when there is maximum horizontal movement of water particles in a landwards direction, and, conversely, on the landwards side as the wave trough passes when the water particles move seawards.

Wave-generated ripples begin to form as soon as the threshold velocity for grain movement is reached. As orbital velocities increase, the height of the ripples decreases until the sediment moves as a suspended 'sheet flow', backwards and forwards over the sea-bed.

Occasionally, the ripple marks found in runnels (Section 5.1.3) are a fascinating combination of straight-crested wave-generated ripples interspersed with asymmetrical current-formed ripples, trending roughly at right angles to the wave ripples (Figure 5.12). The resulting patterns are known as **ladder-back ripples**. Sometimes, the current ripples give an indication of the longshore current direction, but in the case of Figure 5.12 they are probably formed by water draining from the runnel on the ebb tide.

Another common feature you have probably seen on sandy beaches is the rhomboid pattern formed by fast-flowing backwash (Figure 5.13). The origin of these patterns is not fully understood but it may be that minor irregularities in the surface texture of the beach sediment are sufficient to split the backwash into diverging minor currents which cross each other to leave the characteristic rhomboid marks.

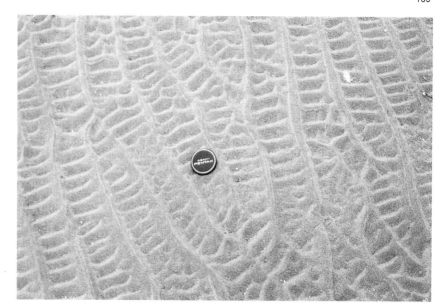

Figure 5.12 Ladder-back ripples: wave-formed ripples with linear, bifurcating crests, in between which are smaller, current-formed ripples running at right angles to the wave-formed ripples. The scale is given by the camera lens cap.

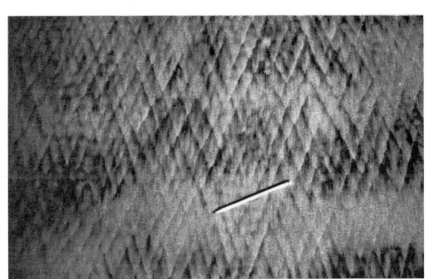

Figure 5.13 Rhomboid marks in beach sands. The scale is given by the pencil.

5.5 SUMMARY OF CHAPTER 5

1   The littoral zone is divided into various component zones according to the influence of tides or waves, or according to the sediment profile. The tidal zones are the backshore, foreshore and shoreface; the wave zones are the swash zone, the surf zone and the breaker zone. The beach profile includes the berm, the beach face with swash bars and runnels, and longshore bars.

2   Water particles in shallow-water waves follow orbital paths which become progressively flattened towards the sea-bed. In shallow water, the maximum orbital velocity u_m and shear stress at the bed increase as the wave height increases and as water depth decreases.

3   Sediment movement begins when u_m reaches a critical threshold value, u_t, for a given grain size. There is no unique set of conditions which determine u_t for a given grain size and it may be achieved from many different combinations of wave height, wave period and water depth. The threshold velocity for a given grain size increases as the wave period increases.

4   Beneath a wave, sediment is moved landwards as the crest passes and seawards as the trough passes. Strong shoreward velocities move coarser sediment landwards as bedload and finer sediment as suspended load. Weaker seawards velocities, of longer duration, move only the finer bedload and suspended load seawards. This leads to a net movement of coarse sediment landwards and fine sediment seawards.

5   Wave-induced longshore currents are generated when waves break obliquely to the shoreline. These currents, and the zig-zag movement of swash and backwash on steep beaches, move sediment along the shoreline. Rip currents are formed when wave crests contain waves of different heights due to differential shallowing of the sea-bed beneath the wave crest. The horizontal pressure gradient between wave set-ups of different heights leads to the flow of water along the shore in opposing directions. Where the flows meet, water is returned seawards as a rip current.

6   The wave power available for longshore sediment transport can be calculated from the wave group speed, significant wave height and the angle the wave crest makes with the shoreline. The rate of sediment transport along the shoreline can be estimated using the wave power.

7   Most net sediment transport occurs when movement by currents is enhanced by wave action. Wave action lifts sediment into suspension where it is transported by currents which, by themselves, may be unable to lift sediment off the sea-bed. The enhancement factor of waves on the shear stress at the bed increases as the current speed decreases and, for currents less than a few cms^{-1}, may be several orders of magnitude.

8   Coarse-grained sediments lead to steep beach profiles because water is readily lost through percolation and the backwash is too weak to move much of the sediment that has been transported up the beach face by the swash. Conversely, fine-grained sediments lead to shallow profiles. Small, gentle waves and swell waves tend to build beaches up and steep, storm waves tend to tear them down and flatten them. The control of grain size is more important than wave steepness on beaches made of coarse sand or gravel-sized particles, but the control of wave steepness is more important on beaches made of fine sand.

9   Straight-crested, symmetrical ripples form as the result of the oscillatory water movement beneath waves. Rhomboid patterns are formed as backwash flows over minor irregularities in the sediment surface.

Now try the following questions to consolidate your understanding of this Chapter.

QUESTION 5.10   Consider Figure 5.10. How would you expect the wave enhancement factor of a wave of given height and period to vary as it moved into shallower water?

QUESTION 5.11 Explain whether large waves or small waves are more likely to generate strong longshore currents.

QUESTION 5.12 Many years ago, improvements were made to the main river channel of the St John's River, Florida, which had silted up so much that it could be used only by small boats. A deep channel was dug for ships and large jetties were built immediately to the north and south of the estuary mouth which extends eastwards into the Atlantic Ocean at Jacksonville. The predominant longshore drift is southwards. Predict the outcome of building these jetties on the sediment budget of the coastline immediately to the north and south of the estuary.

'All the rivers run into the sea; yet the sea is not full.'

Ecclesiastes, I, 7.

Wherever wave power is relatively low along a stretch of coastline, and the tidal range is moderate to large, then **tidal flats**, rather than beaches, are likely to develop. Tidal flats usually have very low gradients, in the order of 1:1000 at the landwards edge, and are composed predominantly of silts and clays, instead of sands. The large tidal range and shallow gradients mean that waves do not break on any one part of the flats for a long time and, consequently, the flood and ebb tidal currents are more effective at sediment transport than is wave action.

QUESTION 6.1   With the aid of an atlas, can you suggest why tidal flats are well developed around the Wash on the east coast of England, along the north coast of the Netherlands and along the coastline of the United Arab Emirates?

There are some exceptions where tidal flats occur on coasts facing the open sea, e.g. Surinam on the north-east coast of South America. In such cases, tidal-flat development is encouraged by a combination of exceptionally high concentrations of fine-grained, suspended sediments in the coastal waters and a very gentle offshore slope. Elsewhere, tidal flats are usually restricted to regions in the shelter of features such as spits (e.g. Orford Ness on the Suffolk coast), barrier islands (like the Friesian Islands), and coastal embayments or estuaries.

In the first part of this Chapter, we look at the processes leading to sediment transport and deposition on tidal flats. In the second part, we turn our attention specifically to estuaries where, although the same general principles apply, the interaction of tidal currents and river flow modifies the pattern of sediment transport.

6.1   SEDIMENT TRANSPORT AND DEPOSITION ON TIDAL FLATS

Tidal flats are flat and almost featureless areas which occur along some stretches of coastline and within estuaries. They are often backed by areas of salt-marsh and dissected by a network of tidal channels (Figure 6.1). Seawater enters the tidal channels on the flood tide, gradually filling them as the tide rises until the water spills over and floods the adjacent flats. After the slack water of high tide, the water drains back off the flats and through the tidal channels until the entire tidal flat is exposed once more. This pattern of water movement, and the interaction between tidal currents and wave action, has important consequences for the transport and distribution of sediment on the tidal flat.

6.1.1   SEDIMENT DISTRIBUTION ON TIDAL FLATS

In the simplest situation (e.g. the tidal flats of the Netherlands), there is a seawards progression in grain size from mud-dominated sediments at the

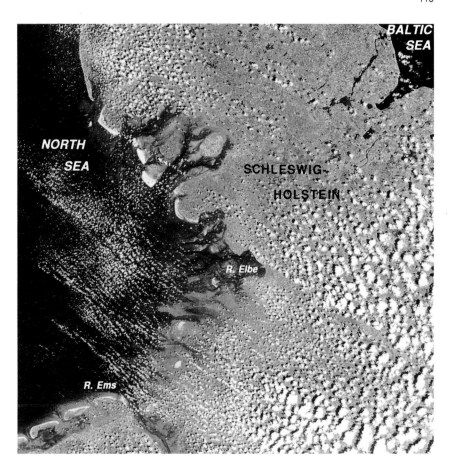

Figure 6.1   Landsat Multispectral Scanner image of part of the German North Sea Coast. Islands and exposed sand bars are shown in white, tidal flats are grey and tidal channels, estuaries and sea areas are black.

landwards end to sand-dominated sediments at the seawards end (Figure 6.2(a)). However, there are departures from this trend (e.g. sediment zonation in the Wash is more complicated due to more extensive wave action in the intertidal zone).

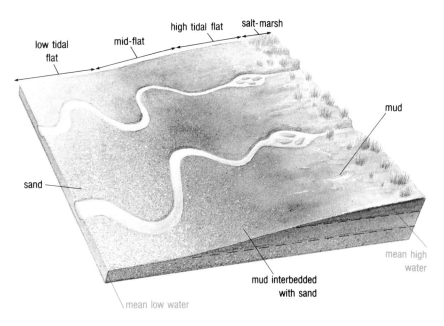

Figure 6.2(a)   A simple model for sediment zonation on tidal flats. Note that the vertical scale has been exaggerated.

Figure 6.2(b) Photograph of a typical estuarine tidal mud-flat.

The **low tidal flats** are submerged for most of the tidal cycle and are subjected to strong tidal currents and minor wave action during this period. Even at the slack water of low and high tide, there is some sediment disturbance by waves. Consequently, muds are kept in suspension and sediments are deposited only from the bedload. These consist of well-winnowed sands.

QUESTION 6.2 What sort of bed forms might you expect to see on the low tidal flat?

The **mid-flats** are submerged and exposed for roughly similar periods. They are usually submerged during the mid-tidal cycle when tidal currents may reach their maximum speeds and so the sediment may be affected by these strong tidal currents, although wave action is very weak. Bedload transport and deposition of sands are again dominant, accompanied by the development of current-formed ripples. However, during the period of slack high water, fine muds held in suspension are able to settle out, forming characteristic mud drapes over the surfaces of the previously formed ripples.

The **high tidal flats** are submerged only at high tide when current speeds fall to zero. No bedload transport or deposition occurs but during the slack water period muds settle out of suspension to form the mud-flats (Figure 6.2(b)). When the tide turns, these muds will be eroded only if the ebb current generates a shear stress at the bed which is greater than the critical shear stress required to erode the sediment. Because muds are cohesive sediments, they are difficult to erode after deposition (Section 4.2.2). In laboratory experiments using muds from the Wadden Sea, currents of $0.4–0.5\,\mathrm{m\,s^{-1}}$ were required to resuspend muds that had been allowed to settle for 16 hours, whereas redeposition did not occur until current speeds had decreased to between 0.1 and $0.2\,\mathrm{m\,s^{-1}}$.

The deposition of fine-grained silts and clays on the high tidal flats is also encouraged by their settling lag (Section 4.4.2). As the flood tide inundates the tidal flats and the current begins to slacken, these grains begin to settle from suspension as soon as the critical depositional shear velocity is reached. However, they do not settle vertically through the water, but are carried landwards as they sink, by the still-moving current. Eventually, they are deposited at some distance inland of the point at which the critical depositional shear velocity was reached. Assuming that the ebb tide and flood tide current speeds are equal, when the tide turns the deposited sediments will not be resuspended until much later in the flow. This effect is enhanced by the cohesive properties of the sediment. As a result, on the ebb tide the sediment grains will be suspended for a shorter period than on the flood tide and will not move back offshore as far as they moved inshore. Thus, the high tidal flat is a zone of rapid sediment accretion, and, as its level is raised with fresh accumulations of mud, the degree and duration of submergence during high water decreases.

Ultimately, the flat is exposed for sufficiently long periods for colonization by land plants to begin. The most common pioneering salt-tolerant plants in Western Europe are *Salicornia* (the fleshy marsh samphire) and *Spartina* (the tough marsh cord grass—Figure 6.3). The plant roots help to bind the sediment and prevent further erosion. More significantly, the plant stems retard the flow, encouraging still further the deposition of silts and clays. Total colonization of the muds of the high tidal flat leads to the development of a salt-marsh, flooded normally only during high spring tides. Thus the salt-marsh (and the other zones of the tidal flat) gradually extends seawards, and the older, landwards regions are flooded less frequently. However, the deep drainage channels persist long after the marsh has become dry land.

(a)

(b)

Figure 6.3(a) The binding of muds by *Spartina* grass roots in a salt-marsh.

(b) The aerial roots of mangrove trees help to trap muds and bind sediments.

6.1.2 LOW-LATITUDE TIDAL FLATS

In humid tropical regions, mud-flats are often colonized by mangrove trees whose aerial root systems tend to trap muds. Consequently, mangrove swamps develop above mean high tide level in place of salt-marsh. Mangrove swamps are particularly extensive on the tidal flats in between the tidal channels of the Niger delta, for example.

In low latitudes, wherever little terrigenous sediment is supplied to coastal regions, carbonate sediments are able to accumulate. Here, the sediments of the intertidal region are dominantly carbonate muds (Section 5.4.1) containing a high proportion of faecal pellets. These muds are colonized by blue–green algae which trap and bind the sediments, in the same way that salt-marsh plants do, to form algal mats. Along arid coastlines, such as that of the United Arab Emirates which borders the Persian Gulf, the seawards accretion of sediments leaves the older areas of algal mat stranded above sea-level. They are subject to intense surface evaporation, particularly after they have been flooded by seawater during occasional storms and extra-high tides.

QUESTION 6.3(a)   If water is lost at the surface of the mats by evaporation, what is likely to happen at their base, which is still below mean sea-level?

(b)   What will happen to the salts dissolved in the seawater?

This process is referred to as 'evaporative pumping'; the environment resulting from the sequence of carbonate sediments and evaporites is known as a **sabkha**, the Arabic word for a salt-flat.

6.2 ESTUARIES

The word estuary is derived from the Latin word *aestus*, meaning tide, and the adjective *aestuarium*, meaning tidal. Most people would recognize an estuary as the region where a river meets an inlet of the sea (Figures 6.4(a) and 6.5). As an everyday definition, this may seem perfectly satisfactory, but it gives no indication of how far up-river an estuary extends or of the interaction between the fresh river water and the saline seawater. In order to answer these points, the following catch-all definition has been proposed:

> An estuary is an inlet of the sea reaching into a river valley as far as the upper limit of tidal rise, usually being divisible into three sectors: (a) a marine or lower estuary, in free connection with the open sea; (b) a middle estuary, subject to strong salt and freshwater mixing; and (c) an upper or fluvial estuary, characterized by freshwater but subject to daily tidal action.
>
> (R. W. Fairbridge (1980) in E. Olausson and I. Cato (eds.) *Chemistry and Biogeochemistry of Estuaries*, John Wiley.)

Figure 6.4(b) illustrates this description.

Most estuaries are geologically very young; they have developed since the latest post-glacial rise in sea-level inundated coastlines and drowned the mouths of river valleys. They are now being progressively infilled with sediment. Today, only those rivers which transport small amounts of sediment and discharge it into coastal waters (where wave and tidal current action are sufficiently strong to disperse the sediment) have open estuaries of the type seen round the British Isles. However, where the sediment discharge is high and there is limited wave and tidal current action, then the open estuary rapidly fills in and a delta grows seawards at the expense of the estuary (see Chapter 7).

6.2.1   ESTUARINE TYPES

Estuaries are far from uniform in character and the differences are mainly due to variations in tidal range and river discharge, which affect the

(a)

Figure 6.4(a) Aerial view of the seawards end of an estuary at low tide. The main river channel hugs the left-hand side (looking landwards).

(b) A schematic map of a typical estuary showing the divisions into lower, middle and upper estuary based on the definition by Fairbridge. The boundaries are transition zones that shift according to the seasons, the weather and the tides.

uppermost
limit of
tide

boundaries are subject
to seasonal shifts,
etc.

salt water, freshwater
mixing

influence
of seawater
dominant

sea

freshwater dominant
but subject to tidal influence

(b)

UPPER ESTUARY   MIDDLE ESTUARY   LOWER ESTUARY

Figure 6.5 Landsat Multispectral Scanner image of the Humber estuary, north-eastern England approximately two hours before low water (taken in 1976 before the building of the Humber Bridge). Seawater and river water are black. The yellow coloration in the estuary is due to high concentrations of suspended sediment and so shows the pattern of sediment distribution and movement in the estuary. Although most of the blue colour represents built-up areas, the blue bands bordering the north bank of the Humber as far as the spit at the mouth, and extending southwards from the estuary mouth along the coast, are intertidal mud.

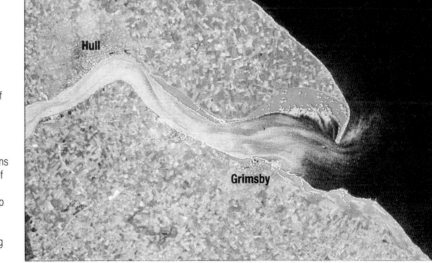

Hull

Grimsby

118

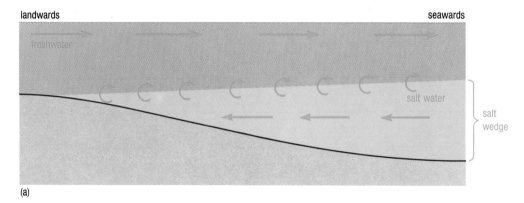

landwards   seawards

freshwater

salt water

salt
wedge

(a)

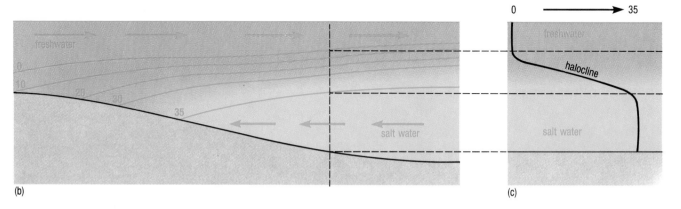

freshwater

salt water

0 ──────► 35

freshwater

halocline

salt water

(b)   (c)

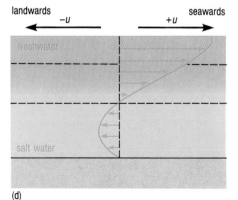

landwards   seawards
        −u   +u

freshwater

salt water

(d)

Figure 6.6   Diagrammatic representation of water circulation, salinity distribution and velocity gradients in a salt wedge estuary.

(a)   Longitudinal profile to show water circulation. The horizontal arrows show the residual circulation. This is seawards at the surface due to mixing and river flow, and landwards at the bottom due to vertical mixing across the river water/seawater interface.

(b)   Longitudinal section of the salinity gradient showing a marked halocline.

(c)   Salinity–depth profile at the position indicated by the dashed vertical line in (b) showing a marked halocline.

(d)   Velocity–depth profile along the dashed vertical line in (b) (longitudinal profile) to show the residual flows.

extent to which saline seawater mixes with fresh river water. On this basis three main types of estuary are recognized: salt wedge, partially mixed and well-mixed estuaries.

Salt wedge estuaries develop where a river discharges into a virtually tideless sea. The less dense river water spreads out over the surface of the denser, saline seawater which, because there is virtually no tidal current movement, may be considered for the time being as a stationary salt wedge penetrating and thinning up-river (Figure 6.6(a)). Between the freshwater and seawater, there are very sharp density and salinity gradients so that a stable **halocline** develops and the two water masses do not mix easily (Figure 6.6(b) and (c)). However, because one layer of water is moving over another, shear stress occurs at the interface, producing turbulence at the base of the freshwater and generating a series of internal waves along the interface (see Section 1.1.1). These can break, ejecting small quantities of salt water into the overlying turbulent freshwater.

QUESTION 6.4   If salt water is lost from the salt wedge into the overlying freshwater with no corresponding gain of freshwater by the salt wedge, what must happen in order to maintain the overall volume of water in the estuary?

The position of the salt wedge is dependent on the river flow. When the discharge is low, the salt wedge can penetrate further inland than when the discharge is high. Only rivers with very low rates of sediment discharge (e.g. the rivers draining Texas and discharging into the Gulf of Mexico) form open salt wedge estuaries. If the discharge of sediment is high, then it tends to accumulate to build a delta. This is the case further east in the Gulf of Mexico where the Mississippi discharges. It also occurs where the Rhône and the Nile enter the Mediterranean Sea; where the Po discharges into the Aegean Sea; and where the Danube discharges into the Black Sea.

Partially mixed estuaries occur where rivers discharge into a sea with a moderate tidal range. Tidal currents are significant, and so the whole water mass moves up and down the estuary with the flood and ebb tides. Consequently, in addition to the current shear at the salt water/freshwater interface, friction at the estuary bed creates shear stress there, and generates turbulence which causes even more effective mixing of the water column than that caused by waves at the freshwater/salt water interface. Not only is salt water mixed upwards, but freshwater is mixed downwards (Figure 6.7(a)). This two-way mixing across the halocline makes it much less well defined (Figure 6.7(b)).

The freshwater flowing seawards is now mixed with a relatively high proportion of salt water so that the compensating landwards flow from the sea is much stronger than in the salt wedge estuary (Figure 6.7(c)). These upper seawards and lower landwards flows distort the isohalines in a longitudinal section of the estuary, as shown in Figure 6.7(a).

The currents caused by the mixing of freshwater and salt water in salt wedge and partially mixed estuaries are referred to as **residual currents** and they are, typically, less than 10% of the magnitude of the tidal currents superimposed on them. However, they are important when we consider how sediment is transported in different types of estuaries.

Towards the head of the estuary, the net landwards bottom flow of water diminishes and the net seawards upper flow increases. The depth at which there is no net landwards or seawards movement of water (where the velocity profile in Figure 6.7(c) crosses the vertical zero velocity line) increases until eventually it coincides with the bed of the estuary. At this point, there is no residual movement of water landwards. This is defined as the **null point** of the estuary.

Would you expect the null point to be fixed in position in a particular estuary?

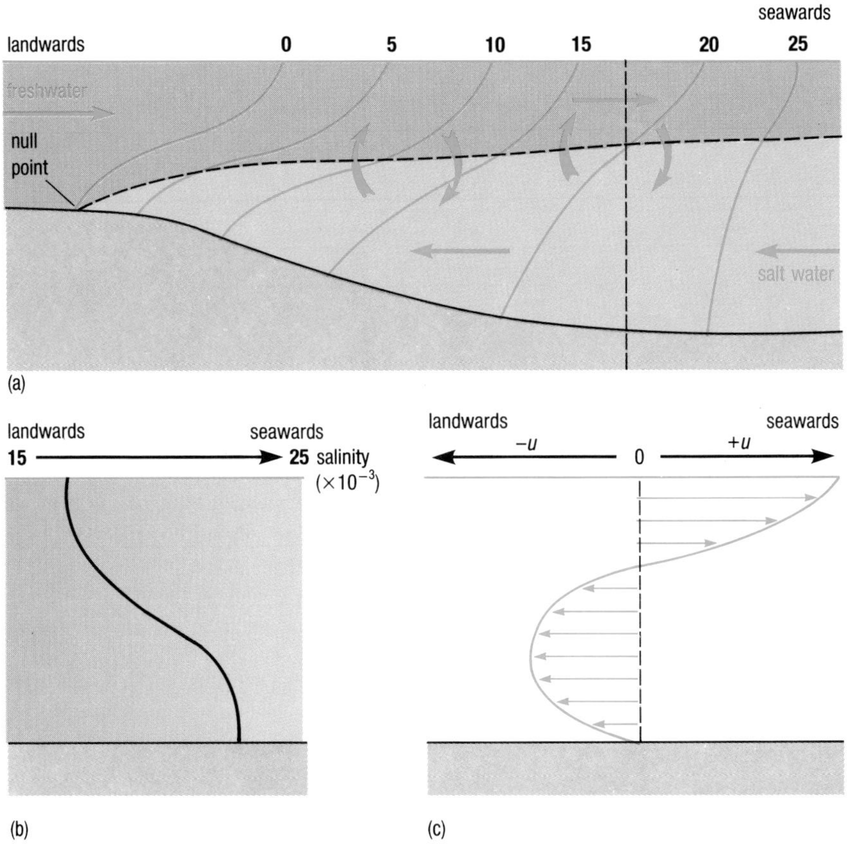

(a)

Figure 6.7 Diagrammatic representation of water circulation, salinity and velocity gradients in a partially mixed estuary.

(a) Longitudinal section to show water circulation and salinity gradient. The dashed sub-horizontal line is the depth at which there is no horizontal residual flow either seawards or landwards.

(b) Salinity–depth profile along the dashed vertical line in (a) showing the poorly defined halocline.

(c) Velocity–depth profile along the dashed vertical line in (a) showing marked upstream residual flow of seawater at the bed.

No. The actual position fluctuates up and down the estuary as the tidal range changes during the fortnightly spring–neap cycle. Also, there will be some seasonal variation: it will move upstream when the river discharge is low, and downstream when the discharge is high. However, there will be an overall long-term mean position. The null point has important consequences for sediment distribution in estuaries and we shall consider this further in Section 6.2.3

Partially mixed estuaries are common on the coasts of north-eastern America (e.g. the James River in Virginia), and north-western Europe (rivers like the Mersey and the Thames).

In broad, shallow, estuaries where the tidal range is high, and the tidal currents are strong relative to the river flow, the water column becomes completely mixed. Such well-mixed estuaries include the Severn estuary, the Firth of Forth in Scotland, the Gironde estuary opening into the Bay of Biscay, the Rio de la Plata, which opens into the South Atlantic and the Humber estuary (Figure 6.5). In these **well-mixed estuaries**, salinity hardly varies with depth at all (Figure 6.8(a)), although it may vary considerably across the width of the estuary. These estuaries are usually shallow and funnel-shaped: wide at the mouth and tapering rapidly inland (Figure 6.5). The Coriolis force tends to swing the incoming tidal flow and the seawards-flowing river water to the right in the Northern Hemisphere and to the left in the Southern Hemisphere. This means that in the Northern Hemisphere the seawater flows up-estuary on the left-hand side (facing downstream), and the river water flows down-estuary on the right-

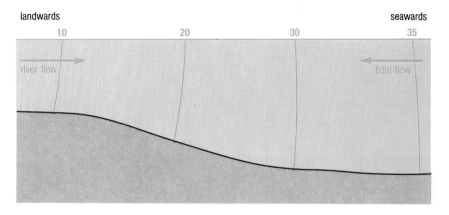

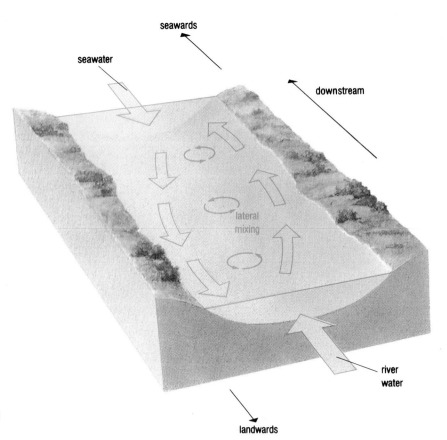

Figure 6.8   Diagrammatic representation of salinity and water circulation in a well-mixed estuary.

(a)   Longitudinal section showing vertical isohalines and the absence of a vertical salinity gradient.

(b)   Block diagram showing lateral deflection of seawater and river water flows due to the Coriolis force, in this case in the Northern Hemisphere. Lateral mixing induces a horizontal residual circulation.

hand side. Some mixing takes place laterally so that a horizontal rather than vertical, residual circulation is developed (Figure 6.8(b)).

A fourth type of estuarine circulation pattern is common in estuaries in arid regions. Very high evaporation rates towards the head of the estuary result in increases in salinity and, therefore, density. This hypersaline water sinks and flows seawards across the estuary floor, and is replaced by a landwards flow of seawater at the surface. This pattern of circulation is referred to as **negative estuarine circulation**, and it is reflected in the distortion of the isohalines seen in Figure 6.9.

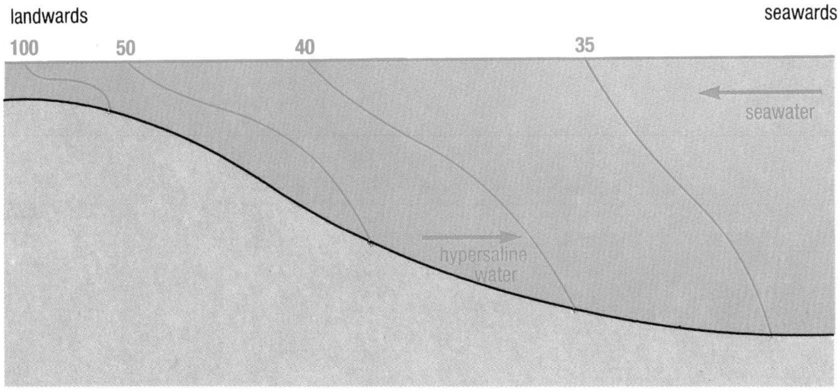

Figure 6.9 Longitudinal section of an estuary with negative circulation to show the residual water circulation and salinity gradient. Note the landwards flow of water at the surface and seawards flow at the bed.

6.2.2 SEDIMENTATION IN ESTUARIES

Much of the sediment that is brought down by rivers is trapped by deposition within estuaries. As you will know if you have ever looked across an estuary at low tide, a large proportion of this sediment is mud.

QUESTION 6.5 Assuming that most deposition occurs at the slack water of high tide, how long would it take clay particles, less than 2μm in diameter to settle through 5m of water to the bed of the estuary? (Settling velocities for different grain sizes are given in Figure 4.9.)

Clearly, some other process must operate in the estuarine environment to encourage the deposition of very fine-grained sediments. This process is the aggregation of the tiny grains to form larger ones which are deposited more rapidly. There are two principal ways in which this may happen: by biological processes and by flocculation.

Biological aggregation results from the ingestion of clay particles by organisms and their subsequent excretion in faecal pellets up to 5mm long with settling velocities measured in centimetres per second, rather than millimetres per hour.

Flocculation occurs as the result of molecular attractive forces known as the van der Waals forces. These forces are not particularly strong, but their strength varies inversely as the square of the distance between two clay particles and they become important when particles are brought very close together. In fresh river water, flocculation does not take place because, for various reasons, the clay minerals normally carry a net negative charge and so the similarly charged clay particles repel one another. In saline waters, such as seawater, interaction with free cations in the water causes a neutralizing effect which reduces the negative charge and allows the molecular attractive force to dominate so that, if clay particles are brought close enough together, flocculation occurs.

Flocculation is an important process in those parts of estuaries where mixing of freshwater and salt water takes place. There are three main ways in which clay particles can be brought close together so that the van der Waals forces can take effect:

1 By turbulence in the water column due to wind action or current shear.

2 By Brownian motion: very small suspended clay particles are shifted about erratically by the random motion of water molecules.

3 By being 'captured' by larger particles which collide with them while settling rapidly through the water column.

Organic substances absorbed by the clays from suspension, and mucous films produced by bacterial activity, have positive charges.

How will these charges affect flocculation?

They will encourage it because unlike charges attract one another.

Although flocculation explains how very fine-grained muds are able to settle in estuaries, it does not explain why such vast quantities of muds are trapped or why there is a net landwards movement of fine sediment where there is an appreciable tidal range in an estuary. For example, between the mid-1960s and the mid-1980s, the Mersey estuary in north-west England is estimated to have accumulated at least 68×10^6 tonnes of sediment that had been transported landwards from the sea (an annual rate of about 3.4×10^6 tonnes). There are three factors that explain the rapid accumulation and landwards movement of sediment.

1   The settling lag of fine-grained sediments together with the cohesiveness of muds. These were discussed in Section 6.1.1 in relation to the rapid accumulation of muds on tidal flats.

2   Tidal asymmetry: as a tidal wave is propagated into an estuary, the wave crest (high water) will travel faster than the wave trough (low water) because the speed of propagation depends on water depth (Section 2.4.3). This results in a slower turn of the tide at high water than at low water, and a longer period of slack water when material may be deposited from suspension.

3   Rapid changes in tidal current speed associated with the flooding and draining of the estuarine mud-flats. As the tide rises, a large volume of water has to flow through the relatively small cross-sectional area of the main tidal channel, and so it must flow with a high speed. At this stage coarse sand, and even gravel, may be moved up and deposited in the channel.

QUESTION 6.6   What will happen when the water spills over the main channel and starts to flood the mud-flats on either side?

The reverse will happen on the ebb tide, and so the velocity curves will be asymmetrical between low tide and high tide. Figure 6.10 shows the general form that the velocity curve would take if the mud-flats were extensive near the low water mark. The maximum flood and ebb currents would occur either side of low water. Where the flats extend to the high water mark, the maximum currents will occur either side of high water.

6.2.3   SEDIMENTATION IN DIFFERENT TYPES OF ESTUARIES

Although the general principles of estuarine sedimentation discussed in the previous Section apply to all estuaries, the patterns of water circulation in different estuary types produce different patterns of sedimentation.

Salt wedge estuaries
The salt wedge estuary is dominated by river flow at the surface with only minor landwards flow of seawater at the bed. Virtually all the suspended material in the estuary is, therefore, river-borne rather than marine in origin. Some of this (usually the coarser material) settles to the bed

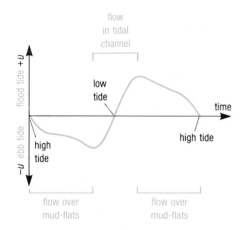

Figure 6.10   The tidal current velocity asymmetry produced during a tidal cycle by the covering and uncovering of intertidal areas near low water mark.

through the halocline and the rest is carried out to sea where flocculation and the reduction in flow as the river currents spread out lead to rapid deposition. As mentioned earlier, if the sediment supply is very large and wave action is weak then a delta may be constructed (see also Chapter 7). At the head of the estuary, where the river meets the salt wedge, the freshwater flows over the salt water leaving the bedload behind, and so a coarse sediment bar may be built close to the tip of the salt wedge.

Partially mixed estuaries

In partially mixed estuaries, the landwards flow of seawater along the bed is sufficiently strong to move sediment up the estuary as far as the null point (Figure 6.7(a)). The material moved may originate as either river-borne sediment which has flocculated on contact with water of increased salinity and settled through the water column, or as marine sediment. Where transport ceases, a **turbidity maximum** is formed where concentrations of suspended sediment around 100–200 p.p.m. (mgl^{-1}) may occur in estuaries with a lower tidal range, but up to 1000–10000 p.p.m. (or $1-10gl^{-1}$) in estuaries with high tidal ranges. The grain size of the suspended sediment is generally less than $10\mu m$. Turbulence at this point and the high concentrations of suspended sediment both encourage flocculation of clays.

The pattern of water circulation illustrated in Figure 6.7(a) does much to maintain the turbidity maximum. Suspended sediment is brought downstream by river transport to the head of the estuary. In the upper estuary, suspended marine sediments, brought up-estuary by the landwards flow of seawater close to the bed, are mixed into the upper layers at the turbidity maximum where the residual flow is down-estuary. A mixture of marine and river-borne sediment is carried back seawards until a point where the mixing of salt water and freshwater is sufficiently reduced to allow the sediment to settle. Some of it is then carried back landwards to the null point in the salt water flow, along with new sediment being brought in from the sea. This form of circulation therefore acts as a sediment trap which retards the escape of sediment to the open sea.

QUESTION 6.7 How would you expect the position of the turbidity maximum and its sediment concentration to vary with the lunar spring–neap cycle?

River flow also affects the position of the turbidity maximum. A very high river discharge can push the turbidity maximum down-estuary, and maybe out of the estuary altogether. However, at very low discharge, the turbidity maximum may be weak and ill-defined.

In some estuaries with a large tidal range, the end product of this cycle is the accumulation during neap tides of fluid mud, which is eroded and resuspended on spring tides (Figure 6.11). It seems that as the tidal amplitude and tidal currents decrease after spring tides, less and less material is capable of being re-eroded and suspended, and more of the suspended load is able to settle from the turbidity maximum to form a layer of fluid mud close to the bed. This effect is enhanced by the longer periods of slack water at the high water neaps than at the high water springs. During the neap tides the fluid mud becomes a little compacted, so that as the tidal range and currents start to increase again, not all of the sediment is re-eroded and some is left permanently deposited.

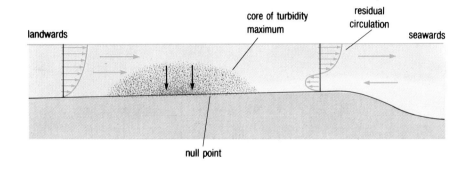

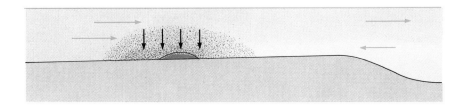

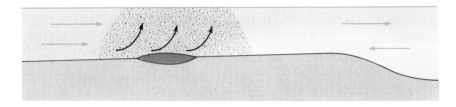

Figure 6.11 The accumulation and movement of fluid mud in a partially mixed estuary. The lens-shaped body of fluid mud extends over distances of 1–10km, and moves backwards and forwards over a few hundred metres with the neap and spring tidal cycle.

Well-mixed estuaries

In the Northern Hemisphere, the flow in a well-mixed estuary (Figure 6.8) leads to the deposition of marine sediments on the left-hand bank, and deposition of river-borne sediments on the right-hand bank facing downstream. In the Southern Hemisphere, river-borne sediments occur on the left-hand bank and marine sediments on the right-hand bank.

Estuaries with negative circulation patterns

These estuaries tend to occur in countries with arid climates where the sediment supply is generally small due to the low rates of weathering. The sediment contains a high proportion of sandy material because of the lack of chemical weathering which produces clays. These sands are deposited at the head of the estuary as bedload. Any finer-grained river-borne sediment is carried seawards in suspension by the residual hypersaline flow just above the bed (Figure 6.9).

6.2.4 PRESERVATION OF THE DYNAMIC BALANCE OF ESTUARIES

The discussion of estuarine circulation patterns and their relationship to estuarine sedimentation (Sections 6.2.1 and 6.2.3) suggests that there is a very delicate balance between the two. Although estuaries make good sites for natural ports and harbours, they are slowly silting up as they adjust to the most recent post-glacial rise in sea-level, and there is always a temptation to interfere with them to keep them open to shipping as long as possible.

QUESTION 6.8 Silting up of the lower Savannah River Estuary (a partially mixed estuary in Florida) was threatening navigation as far upstream as the port of Savannah itself. To combat this, the navigation channel was deepened by dredging which caused a concomitant increase in salt water penetration up the estuary. Can you predict what happened as a result, and why?

A similar situation to this was observed in the Mersey estuary of north-west England following the building of channel walls, started in 1909, to semi-canalize a nine-mile stretch of the main shipping channel in Liverpool Bay. Not only was an increased sediment load brought up-estuary from Liverpool Bay and the Irish Sea, but the sediment load was unable to be dispersed over the sides of the main channel. The overall effect was to encourage sedimentation in the estuary.

6.3 SUMMARY OF CHAPTER 6

1 Tidal flats develop along coastlines where wave action is weak and the tidal range is moderate to large, usually where coastlines are sheltered by islands, or where there are coastal embayments.

2 In general, tidal flat sediments become coarser seawards due to the increasing influence of tidal current action. The low tidal flats are subjected to strong tidal current action for most of the tidal cycle and are comprised of sands; the mid-flats are subjected to alternating strong tidal currents and slack water and so comprise both sands and muds. The high tidal flats are only submerged when tidal currents are weak at slack water, and so are mud-dominated. Colonization of the mud-flats by plants leads to the formation of a salt-marsh. Rapid formation of tidal mud-flats is due to the cohesive nature of muddy sediments, settling lag and the growth of marsh plants.

3 In the humid tropics, mangrove swamps replace salt-marshes. In low latitudes, where terrigenous sediment input is negligible, carbonate muds accumulate, colonized by blue–green algae. Along arid coastlines sabkhas, or salt flats, develop.

4 Estuaries are tidal inlets where mixing of freshwater and seawater occurs. They are ephemeral features, now being infilled with sediment. Four types of estuary are recognized, characterized by their circulation patterns:
(i) Salt wedge estuaries develop in virtually tideless seas and where sediment discharge is low. They involve dominant seawards flow of freshwater at the surface, with minor landwards movement of salt water at the bed. Turbulence at the halocline leads to the transfer of salt water to the freshwater layer. (ii) Partially mixed estuaries develop where there is a moderate tidal range. Mixing of fresh and salt water occurs due to turbulence generated by current shear, both at the bed and the freshwater/seawater interface. There is significant movement of water both seawards at the surface and landwards at the bed. (iii) Well-mixed estuaries develop where the tidal range is high. There is very little variation in salinity with depth but a marked lateral salinity gradient. The mean velocity is seawards at all depths. (iv) Negative estuarine circulation develops where very high evaporation rates at the head of the estuary

lead to the sinking of dense, saline water, and a landwards flow of less saline seawater at the surface to replace it.

5 Fine sediment is deposited by aggregation into larger particles with higher settling velocities. Aggregation occurs either by biological processes involving the ingestion and excretion of sediment as faecal pellets, or by flocculation in saline water. Flocculation is aided by the adsorption of organic substances, mucous films, and turbulence in the water column.

6 Net landwards transport of sediment is enhanced by settling lag, tidal asymmetry, the cohesive properties of fine sediments, and the rapid changes in tidal current speed associated with the flooding and draining of mud-flats. In a salt-wedge estuary, most suspended material is river-borne and is deposited by settling through the halocline. In partially mixed estuaries, river-borne suspended sediment is transported seawards near the surface and marine-derived suspended sediment is transported landwards near the bed, together with any river-borne sediment which has settled through the water column. A turbidity maximum develops near the null point which varies in position according to the state of the tide and river flow. Sometimes, fluid mud forms during neap tides and is dispersed by the following spring tide. In well-mixed estuaries in the Northern Hemisphere, marine sediments are deposited on the left-hand bank (looking seawards), and river-borne sediments on the right-hand bank. This pattern is reversed in the Southern Hemisphere. In estuaries with negative circulation patterns, sandy material is deposited from the bedload at the head of the estuary and fine silts are carried seawards in suspension by hypersaline flow at the bed.

Now try the following questions to consolidate your understanding of this Chapter.

QUESTION 6.9 Figure 6.12 shows the distribution of salinity in Chesapeake Bay, Maryland, on the north-eastern USA coast.

(a) Explain the shape of the surface isohalines in the main estuary.

(b) Describe the pattern of water circulation you would expect to occur in the main estuary.

(c) What pattern of sedimentation would you expect to occur in the main estuary?

(d) Is there any evidence to suggest that the water circulation in the minor estuaries to the west (e.g. the estuaries of the Potomac and James Rivers) is the same as in the main estuary?

QUESTION 6.10 Figure 6.13 shows the variation in tidal current speed and suspended sediment concentrations during a complete semi-diurnal tidal cycle at a recording station towards the landwards end of Chesapeake Bay. The sediment concentrations and current speeds were measured at a depth of six metres, fairly close to the bed.

(a) How do the tidal current speeds close to the bed vary with the tidal cycle?

(b) A similar pattern is seen in tidal current speeds measured at the surface, except that the maximum speeds of the two ebb tides are approximately the same. What are the implications of this observation,

Figure 6.12 Map of the distribution of surface salinity (in parts per thousand) in Chesapeake Bay, Maryland, north-eastern coast of the USA. (For use with Question 6.9.)

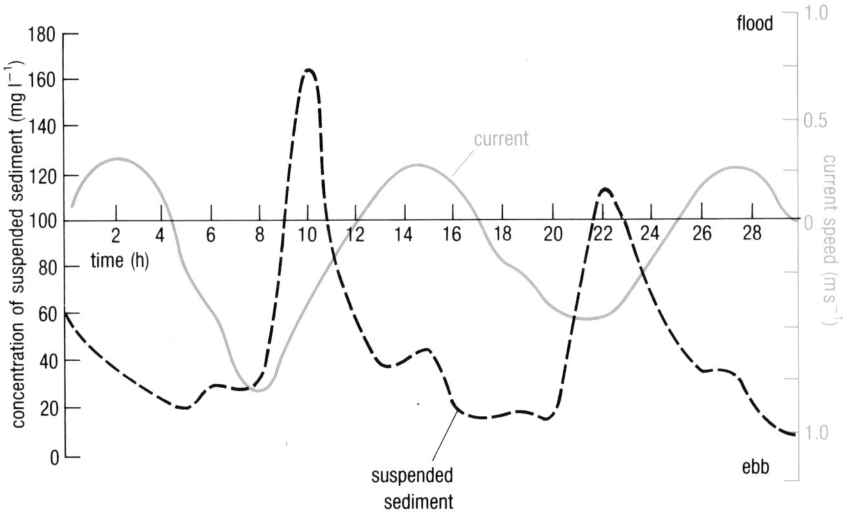

Figure 6.13 Variations in tidal current velocity and suspended sediment concentration during a semi-diurnal tidal cycle at 6m depth from the surface at a recording station in Chesapeake Bay. (For use with Question 6.10.)

and your answer to (a), for the variations in shear stress at the bed during the tidal cycle?

(c)   How does the concentration of suspended sediment close to the bed relate to the curve for current speeds, and your predicted variations in shear stress at the bed?

CHAPTER 7   DELTAS

'...then sands begin
To hem his watery march, and dam his streams,
And split his currents; ...
...till at last
The long'd-for dash of waves is heard...'

From *Sohrab and Rustrum* by Matthew Arnold.

When the sediment discharge from a river is so large that former estuaries become completely filled, and wave and tidal current action are unable to disperse the sediment that reaches the river mouth, then it is deposited seawards of the mouth in the form of a delta. In this context, a **delta** is a coastal accumulation of river-borne sediment extending both above and below sea-level close to a river mouth. However, deltas also develop in freshwater lakes and inland seas—in fact, in any body of water where sediment can accumulate faster than it can be dispersed.

The term 'delta' was first used by the Greek historian, Herodotus, around 450 BC to describe the triangular accumulation of sediments at the mouth of the Nile (Figure 7.1) which, being triangular, resembled the Greek capital letter delta, Δ. As you will see later, deltas only develop this particular shape when a certain combination of wave and riverine (or fluvial) influence occurs. In fact, a wide variety of delta types occurs depending on the relative influences of the river, wave action and tidal currents.

Figure 7.1 The Nile Delta, photographed from the Gemini IV spacecraft.

130

Rivers containing sediment loads sufficiently large to form a delta usually have a large drainage basin and are fed by many tributaries which supply both water and sediment. Precipitation and erosion within the drainage basin (functions of climate, local geology and relief) are critical factors determining water and sediment discharge, and hence they determine whether a delta is likely to develop.

Most of the world's major deltas occur at mid- to low-latitudes (Figure 7.2) and so they often form extensive wetlands of high biological productivity and fertility which, among other things, makes them important conservation areas. They are also regions where thick piles of sediment and vegetation accumulate rapidly and so ancient deltaic sediments are important as source rocks for deposits of oil, gas and coal.

Figure 7.2 The location of some important deltas in relation to the world's largest drainage basins and coastal tidal ranges. Where deltas are shown by circles, the size of the circle is proportional to the suspended sediment load. The arrows show coastlines where storm waves are most active.

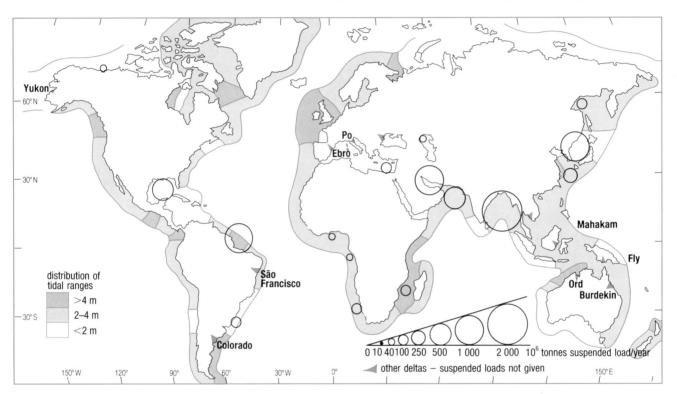

7.1 THE STRUCTURE OF A DELTA

An aerial view of a coastal delta (e.g. Figures 7.1, 7.3, 7.4, 7.5) shows an extensive lowland area above sea-level, usually crossed by a network of active and abandoned channels which are separated by either vegetated or shallow water areas. This is the region known as the **delta plain**. Major delta complexes like the Mississippi, the Amazon and the Ganges–Brahmaputra are as much as 200–300 km across and occupy areas as large as Wales or Northern Ireland. The numerous channels are known as distributaries; as a channel becomes blocked with sediment, the flow will split to find new routes round the obstruction, so forming new channels.

Seawards of the delta plain lies the **delta front** which comprises the shoreline and the part of the delta below sea-level where the deltaic sediments dip seawards. This is the part of the delta where the

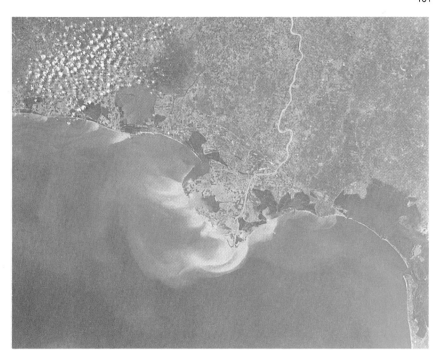

Figure 7.3 The Po Delta in the Adriatic in October, 1984, showing extensive sediment plume.

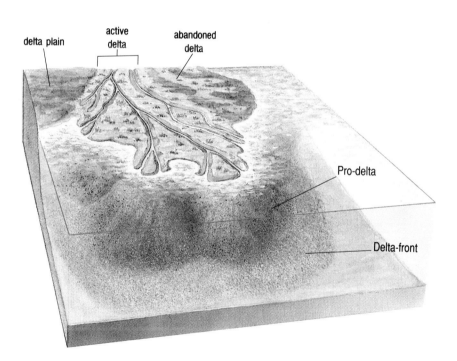

Figure 7.4 The structure of a delta.

fluvial bedload is deposited and so it consists mainly of sandy sediments. The deeper, offshore zone of the delta is the **prodelta**, which receives much of the silt and clay that is transported seawards in suspension. In most delta systems, the prodelta merges imperceptibly into a normal marine shelf-sediment environment. However, when the input of terrigenous sediment is very high (e.g. the Ganges–Brahmaputra delta), terrigenous sediment cascades down the continental slope to feed vast submarine fans.

7.2 MIXING AND SEDIMENT DEPOSITION AT DISTRIBUTARY MOUTHS

Although we have drawn a distinction between estuaries and deltas on the basis of sediment supply and deposition, the processes of mixing between seawater and river water that occur in association with deltas are much the same as those described for estuaries in Chapter 6. Differences in the type and degree of mixing at a distributary mouth lead to differences in the patterns of sediment deposition. Differences in the relative influences of river, tidal current and wave action lead, in turn, to differences in the way in which the sediment is redistributed to give a delta its characteristic shape.

7.2.1 RIVER-DOMINATED DELTAS

River-dominated deltas occur where the tidal range is very low and tidal current action is very weak. The best-known and best-described example is the Mississippi delta (Figure 7.5) which has formed where the Mississippi discharges into the Gulf of Mexico. The tidal range in the Gulf of Mexico is less than 0.5 m. Further west, the rivers which discharge into the Gulf of Mexico from Texas carry far less sediment than the Mississippi and so form open estuaries instead of deltas.

Figure 7.5   A Seasat synthetic aperture radar (SAR) image of the Mississippi delta. The long, thin curving features seen in the waters to the east of the delta are surface waves produced where the outflowing freshwater from the Mississippi interacts with the seawater in the Gulf of Mexico.

QUESTION 7.1   Briefly describe the patterns of water circulation and mixing that are likely to occur when a distributary discharges into the sea in a region where tidal current action is insignificant.

Whether or not this density stratification actually occurs and a salt wedge forms (as described in Question 7.1), depends upon the speed of the river water, and the depth of the distributary mouth. The presence of a salt wedge and density stratification affect the way in which sediment is deposited at the distributary mouth.

Density stratification

This is most likely to occur where the speed of the river flow is moderate to low and the distributary mouth is relatively deep allowing the salt wedge to penetrate up-river. As the river water flows from the confines of the distributary mouth into the open sea, it spreads out over the surface of the seawater as a two-dimensional jet (or plume) (Figure 7.6(a)). Such plumes extend several kilometres from the active distributary mouths of the Mississippi (see Figure 7.5). Mixing occurs both at the base of the freshwater where it flows over the salt water, and at the sides of the plume.

QUESTION 7.2(a)   What will happen to the freshwater and salt water at these boundaries where shear stresses occur?

(b)   What will happen to the speed of the freshwater flow as a result?

(c)   What will happen to the coarser river-borne sediment?

When a river carries a high proportion of coarse-grained sediment, the deposition of this sediment at the distributary mouth leads to a shallowing of the mouth and mixing of river water and seawater, rather than density stratification. The Mississippi, however, has an exceptionally fine-grained load. Only 2% is sand-sized, the remainder being silt (48%) and clay (50%). Although some of the finer sediment is trapped within the distributary mouth by normal estuarine processes of sediment deposition (as described in Section 6.1.1), most is carried further seawards to be deposited on the prodelta.

QUESTION 7.3   As sediment-laden freshwater mixes with seawater, what other process will aid the rapid deposition of the fine-grained suspension load on the prodelta?

The lateral expansion of the plume as it moves out of the distributary mouth, and the mixing of freshwater and salt water at the lateral boundaries of the freshwater plume causes a secondary flow system to be set up which helps to modify the pattern of sediment deposition (Figure 7.6(d)). Because the less dense freshwater rests on top of the denser salt water, it is slightly elevated relative to the surrounding seawater and so will tend to flow laterally (as shown in Figure 7.6(d)) creating a zone of **divergence**. At the sides of the plume, where seawater is mixed into the freshwater, seawater will tend to flow laterally towards the plume to replace that lost by mixing. Where the zone of **convergence** or **front** between the freshwater and seawater occurs, water sinks and moves back towards the centre point beneath the plume. Here, freshwater rises again as the result of divergence at the surface. The double circulation cell shown in Figure 7.6(d) therefore leads to divergence at the surface and convergence at the bed. Flow convergence at the bed prevents the lateral redistribution of coarser sediments and so these are confined to a linear path seawards from the distributary mouth.

Close to the channel mouth, linear subaqueous levées, or raised banks of sediment, are formed, which diverge very little as they are built seawards. Consequently, the deltaic distributaries and their deposits tend to be long, straight and finger-like (see, for example, the Mississippi delta, Figure 7.5) to produce the classic 'bird's-foot' appearance.

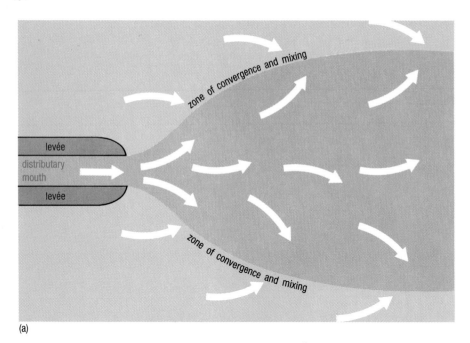

(a)

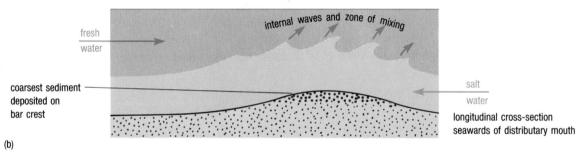

fresh water

internal waves and zone of mixing

coarsest sediment deposited on bar crest

salt water

longitudinal cross-section seawards of distributary mouth

(b)

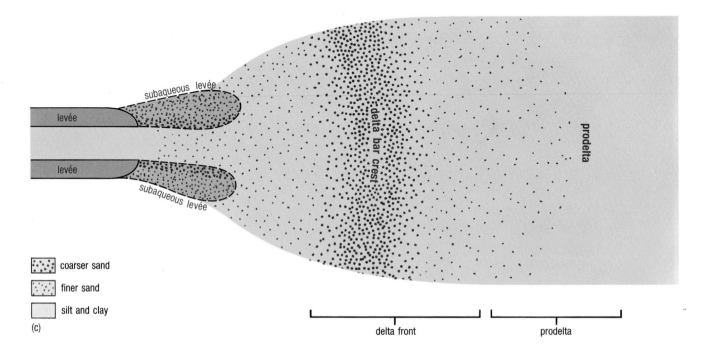

levée

subaqueous levée

distributary mouth

levée

subaqueous levée

delta bar crest

prodelta

coarser sand

finer sand

silt and clay

delta front

prodelta

(c)

Figure 7.6 The patterns of spreading and mixing of freshwater and seawater, and the pattern of sediment deposition at a distributary mouth where density stratification occurs. Arrows show direction of water movement.

(a)   The lateral spreading of a freshwater plume entering a relatively deep salt water basin.

(b)   Mixing and the entrainment of salt water due to the generation and breaking of internal waves at the freshwater–salt water boundary, and deposition of sediment as a delta bar.

(c)   The deposition of sediment close to a distributary mouth.

(d)   Cross-section through the secondary flow system resulting from the lateral flow of the freshwater, and mixing with salt water.

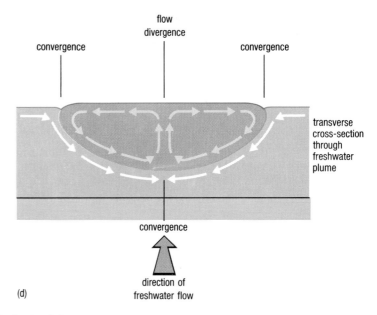

(d)

Turbulent mixing

When the speed of the fluvial discharge is high, then the outflow is turbulent; vigorous mixing occurs with the seawater and so density stratification cannot occur. In the case of the River Amazon, the river outflow is so powerful that it literally forces the salt water back, seawards of the delta bar crest. If the high speed river water is discharged into moderately deep water, then turbulent mixing occurs in three dimensions and the plume can expand both vertically and laterally. However, because of the vertical expansion, the amount of lateral expansion is reduced and the angle of spreading is relatively small (Figure 7.7(a)). As the water is deep, mixing does not occur right down to the bed which is overlain by a layer of unmixed seawater (Figure 7.7(b)).

Would you expect any residual flow (Section 6.2.1) in this seawater layer?

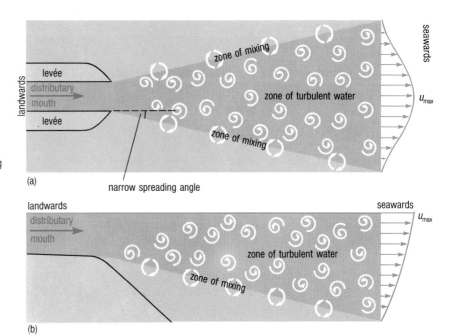

Figure 7.7 The patterns of spreading, turbulent mixing and flow deceleration which occur when the river discharge is high, and into deep water. The length of the blue arrows is proportional to the velocity of the freshwater flow; u_{max} is the maximum flow velocity.

(a)   Schematic plan view of the distributary mouth to show the narrow spreading angle and the horizontal variation of velocity at the seawards edge of the plume.

(b)   Schematic cross-section of (a) to show vertical spreading and turbulent mixing of freshwater and salt water, and the associated vertical velocity profile at the seawards end of the plume.

136

Figure 7.8 The patterns of spreading, turbulent mixing and flow deceleration which occur when the river discharge is high, but into shallow water. The length of the blue arrows is proportional to the freshwater flow velocity and u_{max} is the maximum flow velocity.

(a) Schematic plan view of the distributary mouth to show the wide spreading angle and the freshwater flow velocities at the seawards edge of the plume.

(b) Schematic cross-section of (a) to show turbulent mixing occurring down to the bed. The two velocity profiles show the rapid deceleration of the freshwater flow.

(c) As the freshwater flow decelerates, deposition occurs rapidly, blocking the distributary mouth. The turbulent discharge therefore bifurcates isolating a sediment bar between two new channels and their associated subaqueous levées.

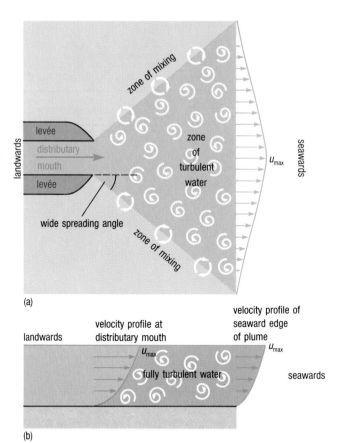

(a)

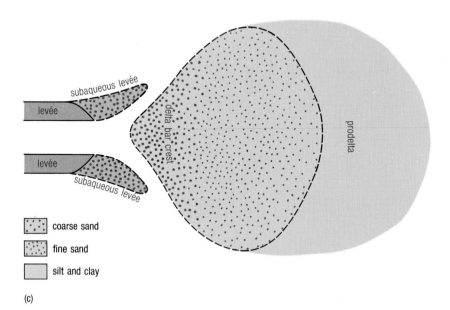

(b)

(c)

Yes, because of vertical mixing. Seawater must move landwards to replace that lost by mixing into the seawards-moving freshwater. However, the shear stress at the bed due to this residual flow is not very great. Deceleration of the freshwater flow is mainly due to turbulent mixing and is nevertheless generally sufficient for sediment to be deposited. As the lateral spread of the flow is restricted close to the outlet, sediment is distributed, once again, over quite a narrow zone.

The discharge from the Mississippi shows interesting seasonal variations. During most of the year, the total flow is around $8400\,\text{m}^3\text{s}^{-1}$. A well-defined salt wedge is able to penetrate upstream and density stratification occurs. However, for between one and three months a year, there is flooding which increases the discharge more than three-fold. The density stratification breaks down and turbulent mixing occurs instead. Because of the associated increase in sediment discharge, there is rapid seawards growth of the deltaic deposits.

Most rivers carry a higher proportion of coarse-grained sediment than the Mississippi, and this is generally deposited close to the distributary mouth, raising the level of the sea-bed. Consequently, it is more usual for freshwater to be discharged into shallow water.

QUESTION 7.4 If high speed river water is discharged into a shallow basin:

(a) What effect will the shallowness of the basin have on the potential for both vertical and lateral expansion of the freshwater plume?

(b) What will be the effect on vertical mixing?

(c) What effect will the answer to (b) have on shear stress at the bed just seawards of the distributary mouth and on the nature of the bedload?

The greater lateral expansion (Figure 7.8(a)) and mixing down to the bed (Figure 7.8(b)) lead to rapid deceleration of the flow (Figure 7.8(a) and 7.8(b)) and so rapid deposition of the bedload (Figure 7.8(c)). Thus, a positive feedback mechanism is generated between sedimentation and flow deceleration. Sedimentation further decreases the water depth, which leads to increased lateral expansion, mixing and deceleration of the flow.

What would be the logical outcome of this situation if flow deceleration and increased sedimentation were allowed to proceed unchecked?

Most probably the distributary mouth would become completely blocked by sediment. In practice, the system usually adjusts as divergent, bifurcating channels are established around the deposited sediment. The flow is now shared between channels and so the extent of both vertical mixing and lateral expansion are reduced, and shear stress at the bed is reduced, too (Figure 7.8(c)).

7.2.2 TIDE-DOMINATED DELTAS

Tide-dominated deltas occur in regions where wave action is limited and tidal ranges are generally in excess of 4m, generating strong tidal currents. For example, near the mouth of the River Ord, on the north coast of Western Australia, tidal currents of more than $3\,\text{ms}^{-1}$ occur during the spring tides. Currents like these have a major effect on the mixing of river water and seawater and on sediment redistribution.

QUESTION 7.5 Predict the effects that a large tidal range and strong currents will have on the following, giving the reasons for your answers.

(a)   Density stratification at the mouths of distributaries, and the flow of water within the distributary mouths.

(b)   Sediment movement within, and close to, the mouths of distributaries.

(c)   The position of the interface between fluvial and marine processes.

Sediments brought down towards a distributary mouth by river flow are rapidly reworked by the tidal currents into a series of linear subaqueous ridges within the distributary mouth, and further seawards. These ridges may be several kilometres long, a few hundred metres wide and may be over 20m high. As the delta builds gradually seawards, former sand ridges are exposed above sea-level and become colonized by vegetation to form linear islands. A series of linear islands and sand ridges projects over 90km offshore in the Ganges–Brahmaputra delta (Figure 7.9). A similar

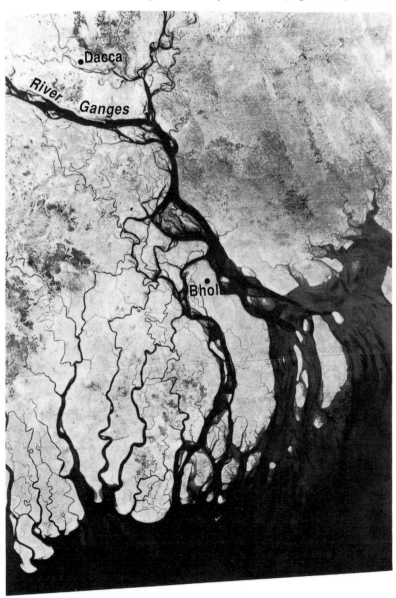

Figure 7.9   Mosaic of two Landsat images of the eastern active Ganges–Brahmaputra delta, showing the ragged outline and funnel-shaped distributary mouths of a tide-dominated delta and the linear sand ridges formed by tidal currents. This view is approximately 180km from side to side.

pattern of sediment movement can be seen from the sediment concentrations in the well-mixed Humber estuary (Figure 6.5). However, the rate of sediment supply by the Humber is too low to permit the formation of linear sand ridges.

As the distributary mouths of tide-dominated deltas are, effectively, well-mixed estuaries, what would you expect their shape to be?

They are typically funnel-shaped. However, the overall outline of the delta complex is very ragged in appearance (Figure 7.9).

7.2.3 WAVE-DOMINATED DELTAS

When a river discharges into the sea where wave energy is high, the delta becomes **wave-dominated**. The effect is very much the same as when waves propagate upstream against an ebbing tide in an estuary (Section 1.5.1). The river flow moving seawards behaves as a current flowing counter to the direction of wave propagation.

QUESTION 7.6 Under these conditions, can you recall what will happen to (a) the speed, (b) the wavelength, and (c) the height of a wave crest?

As a result of these changes, waves approaching the river mouth are liable to break earlier in deeper water than they would normally, and this promotes extensive mixing of seawater and freshwater, and the breakdown of density stratification. Because part of the wave crest impinges on the freshwater plume, it is retarded relative to those parts on either side. The waves are refracted around the plume in a manner analogous to that described in Section 1.4.4 and Figure 1.17, except that the wave crest is slowed down partly because of the influence of the opposing freshwater flow, and not simply because of a shallowing of the sea-floor. Refraction concentrates the wave power on the freshwater plume, enhancing the mixing process still further (Figure 7.10).

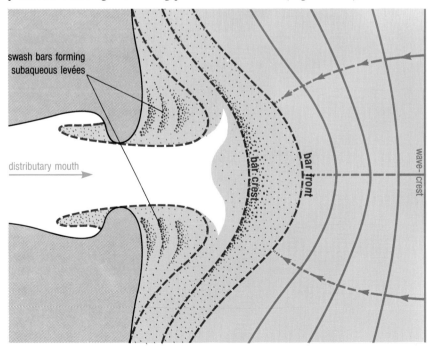

Figure 7.10 The pattern of sediment deposition and wave refraction characteristic of a wave-dominated delta. The dashed blue lines show the direction of wave propagation and the way in which wave energy is concentrated onto the sediment bar, leading to reworking.

Vigorous mixing of seawater with river water leads to the rapid deceleration of the freshwater flow, and equally rapid deposition of sediment. For example, observations of flow rates at the mouth of the Shoalhaven River on the south-east coast of Australia show that when the river is in flood, flow speeds are greater than $2\,\mathrm{m\,s^{-1}}$. However, breaking waves cause immediate mixing of the freshwater and salt water, and within a distance of just over half a kilometre the speed drops to around $0.3\,\mathrm{m\,s^{-1}}$. Only the very fine sediment escapes deposition and is carried seawards to be deposited further offshore. The coarser sediments are deposited in the zone of mixing as a crescentic bar. However, the bar is reworked rapidly by waves and the bedload is moved back landwards by wave action and often forms a series of swash bars (Sections 5.1.3 and 5.2.2—see also Figure 7.10).

The wave-dominated delta shoreline is characterized by straight, sandy beaches and there is usually only a slight protuberance where a distributary mouth meets the sea (Figure 7.11). There are fewer distributaries than in river- and tide-dominated deltas, but occasionally reworked sand being returned landwards blocks a distributary mouth and so the distributary must enter the sea by a new outlet. As the delta grows seawards, the delta plain consists of a whole series of abandoned beaches, now stranded above sea-level.

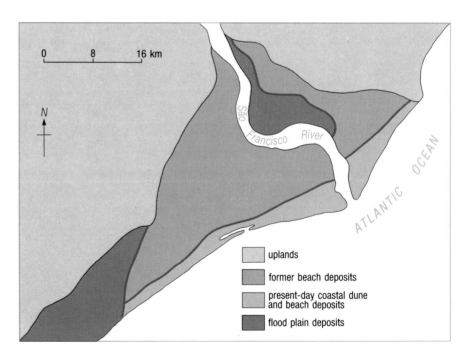

Figure 7.11 The simplified morphology of the São Francisco delta of Brazil, a wave-dominated delta.

7.2.4 OTHER TYPES OF DELTA

Few deltas fall neatly into one of the three types discussed above. Usually, more than one process actively influences the delta shape so that there is a range of delta types intermediate between river-dominated, tide-dominated and wave-dominated. Using qualitative estimates of the relative importance of fluvial, tidal and wave processes, a comparison between deltas can be made by plotting the position of each on a simple triangular diagram (Figure 7.12).

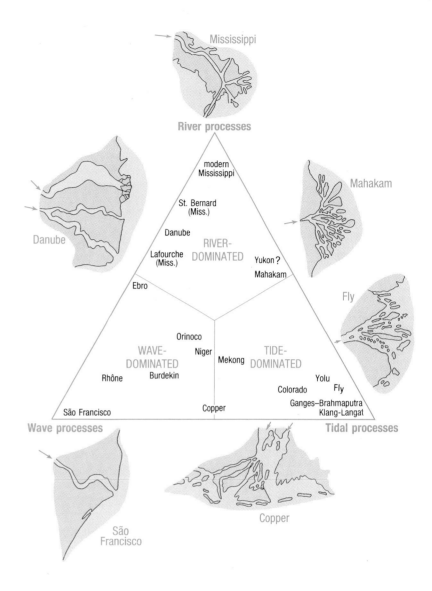

Figure 7.12 The classification of various delta systems based on the relative intensities of river, wave and tidal processes. (Not to scale.)

7.3 SUMMARY OF CHAPTER 7

1   Deltas are coastal accumulations of river-borne sediments which accrete when sediment discharge is so high that it cannot be dispersed by tidal currents and wave action. The main components of the delta are the subaerial delta plain, the shallow water and shoreline region of the delta front and the deeper water prodelta. The shapes of deltas are controlled by the interaction of fluvial, tidal and wave processes.

2   Where both tidal currents and wave action are weak, the speed of the river water is moderate to low, and where the river discharges into relatively deep water, density stratification is set up. Mixing occurs at the salt water/freshwater interface leading to flow deceleration and sediment deposition on a delta bar. Lateral expansion of the flow, and mixing, generate a secondary circulation cell which prevents the lateral dispersion of coarse sediments and encourages the formation of near parallel

subaqueous levées and finger-like deltaic deposits characteristic of the river-dominated delta.

3   In river-dominated systems where the speed of the river water is high, outflow is turbulent and so turbulent mixing occurs. When the river is discharged into deep water, three-dimensional flow, and expansion, and turbulent mixing, lead to flow deceleration and sediment deposition. Sediment dispersal and deposition are limited to a narrow zone because of the small angle at which lateral flow expansion occurs. In shallow water, turbulent mixing occurs down to the sea-bed leading to significant shear stress at the bed close to the distributary mouth, and large bedload transport. However, great lateral expansion and mixing to the bed lead to rapid flow deceleration, rapid deposition and greater lateral sediment dispersion.

4   Where a delta is tide-dominated strong tidal current action prevents density stratification from occurring and the distributary mouth becomes well-mixed. Two-way sediment movement occurs with the formation of sediment ridges within the distributary mouth, parallel to the direction of river flow, producing a ragged outline to the delta.

5   Wave-dominated deltas occur where wave energy is high. The outflowing freshwater behaves as a counter-current, slowing down the approaching wave crests and causing waves to break in deeper water than normal. Waves are also refracted so that wave energy is concentrated on the freshwater plume. Both processes lead to vigorous mixing of freshwater and salt water, rapid deceleration of the freshwater flow and deposition of sediments. Wave action reworks the sediments and moves the coarser sediments landwards to form swash bars and beaches, creating a straight shoreline with only a minor protuberance at the distributary mouth.

6   The shapes of most deltas show the influence of more than one process (fluvial, tidal or wave action).

Now try the following questions to consolidate your understanding of this Chapter.

QUESTION 7.7(a)   Which of the combinations of conditions listed below is (are) most likely to lead to density stratification and salt wedge formation at a distributary mouth?

(i) High river discharge; weak tidal current and wave action; a shallow-water basin.

(ii)   High river discharge; weak tidal currents and wave action; a deep-water basin.

(iii)   Low to moderate river discharge; weak tidal currents and wave action; a shallow-water basin.

(iv)   Low to moderate river discharge; weak tidal currents and wave action; a deep-water basin.

(b)   Which of the conditions (i)–(iv) is (are) most likely to lead to a fully mixed distributary mouth?

QUESTION 7.8   Examine the sketched outlines of the Copper, Fly and Mahakam deltas on Figure 7.12. What evidence is there from the delta shapes to justify their classification on Figure 7.12?

QUESTION 7.9   Examine Figures 7.1 and 7.3 (views of the Nile and Po deltas).

(a)   Explain whereabouts on Figure 7.12 you would plot the Nile delta.

(b)   Explain whereabouts on Figure 7.12 you would plot the Po delta.

CHAPTER 8

SHELF SEAS

'Illimitable ocean, without bound,
Without dimension, where length, breadth and highth,
And time and place are lost.'

From *Paradise Lost* by John Milton.

In this Chapter we are interested in the processes that affect the sea and the sea-bed seawards from the littoral zone to the edge of the continental shelf. It is in this context that the term 'shelf' or 'shelf sea' is used.

Strictly defined, continental shelves are submerged continental margins which slope very gently seawards from the littoral zone, generally to depths of about 100 to 250m at the shelf break. Here, the angle of the continental slope steepens abruptly down into the deep ocean basins. However, in this Chapter we shall be concerned as well with **epeiric seas**, or **epicontinental platforms**, which are flooded areas within a continent, e.g. Hudson Bay, the Gulf of Carpentaria, the North Sea and the English Channel. The same processes operate in these seas as on the true continental shelves and, indeed, much of what we know about shelf processes has been derived from work carried out in the epeiric seas around the British Isles.

Continental shelves vary in size according to their tectonic setting. Along passive (aseismic) margins they tend to be broad, while along active (seismic) margins they are much narrower, although they may be quite extensive behind island arcs, e.g. the East China Sea.

8.1 SHELF SEDIMENTS

To a large extent the nature of shelf sediments is influenced by climate, which controls the way in which adjacent land masses are weathered, and thus the sorts of sediments which are brought down to the ocean margins (Section 3.1). In the humid tropics and subtropics, *chemical weathering* is dominant. This leads to the decomposition of many rock-forming minerals and the formation of clays, and therefore muddy sediments are widespread. In high latitudes, *physical weathering* is dominant, which results in the mechanical shattering of rocks, rather than their decomposition.

QUESTION 8.1(a)   In terms of grain size, what sort of material would you expect to be common in shelf sediments at high latitudes?

(b)   Bearing in mind earlier discussions in Sections 3.1 and 5.4.1, what would you expect the composition of shelf sediment to be at low latitudes where the supply of terrigenous sediment is scarce?

Carbonate sediments are not restricted totally to low latitudes. They will form anywhere if the supply of terrigenous sediment is negligible, although such occurrences are rare. For example, beach and bay sediments at John o' Groats, Scotland, and Sitka Sound, Alaska, contain more than 90% and 66% of carbonate material, respectively. These nontropical carbonates consist mainly of sand- and gravel-sized fragments of molluscs, barnacles and coralline algae. Corals, green algae and inorganic deposits (see Figure 8.1), which provide carbonates in the tropics, are generally absent.

On most shelves, the pattern of sedimentation in relation to present-day climate is complicated by the presence of relict sediments (Section 3.2). You should recall that these are sediments that were deposited in a

(a)

Figure 8.1(a) Various species of the green alga, *Halimeda*, found in the Caribbean. Note the holdfasts at the bases of some of the specimens which anchor the plants in the sediment (approximately half actual size).

(b) Inorganic grains of CaCO$_3$ (ooliths) which are generally 0.2–0.6mm in diameter. These are formed by the precipitation of CaCO$_3$ from seawater around a mineral grain or shell fragment which acts as a nucleus (magnification × 10).

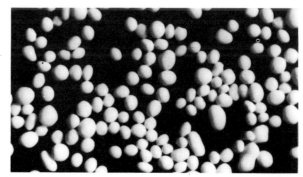

(b)

terrestrial environment when large areas of continental shelves were exposed during the sea-level minima associated with glaciations. Because many of these relict sediments were originally deposited by rivers or ice-sheets, they often contain a mixture of gravels, sands and muds. Like all shelf sediments, they are subject to constant reworking by currents and waves. The deposits that remain after reworking are termed *residual sediments*. A common example is the residual gravel left behind after a poorly sorted glacial deposit (i.e. with a wide range of grain sizes) has been reworked, and the finer-grained material winnowed out.

QUESTION 8.2   Refer back to Section 4.1.3. Can you explain why a current of a given speed may winnow fine-grained material from poorly sorted gravels and deposit it on a smooth, fine-grained bed some distance away even though it has not decreased in speed?

8.2 SHELF PROCESSES

When pipelines or oil rigs are placed on the shelf sea-bed, sediment is scoured as the result of changes to the water flow. In order to predict the extent of scouring, it is important to know the speed and direction of current flow, and the resulting sediment movement on the shelf. Although tidal currents and storm wave action are the most important types of water movement affecting the sea-bed, oceanic currents, local wind-driven currents and storm surges may also cause sediment movement on continental shelves. In this Section we will first deal briefly with the effects of ocean currents and wind-driven currents in transporting sediment.

8.2.1   OCEAN CURRENTS

Although the major ocean currents lie on the oceanwards side of the shelf break, they are able to influence the outer parts of some continental shelves where large eddies spin off the main current and onto the shelf. In the case of the Gulf Stream and the Florida Current, there is also some occasional lateral migration of the current itself onto the shelf. For example, during the summer months, the Florida Current encroaches onto the shelf of eastern North America, whereas during winter it moves offshore. The Agulhas Current of the Western Indian Ocean, however, produces persistent unidirectional flow along the outer shelf of Southern Africa. Such currents are generally a few centimetres per second, capable of transporting fine suspended sediment; however, they may be sufficiently high to cause the erosion of the Blake Plateau by the Florida Current and produce megaripples on the outer Saharan continental shelf by the Canary Current.

8.2.2   LOCAL WIND-DRIVEN CURRENTS

When the wind blows over the sea, it generates a shear stress (known as the **wind stress**) at the sea-surface. This is directly analogous to the shear stress caused by water flowing over the sea-bed (Section 4.1.1). The surface water moves in response to the wind shear stress, and motion is transmitted to successively deeper layers of water. Because of the Coriolis effect due to the Earth's rotation, at the surface the current will deviate from the wind direction by approximately 45°.

Do you recall in which direction the deviation will occur?

It will be to the right in the Northern Hemisphere, and to the left in the Southern Hemisphere. For this reason, longshore winds can lead to offshore transport of surface water and upwelling (Figure 8.2(a)), or onshore transport and downwelling (Figure 8.2(b)), depending on wind direction, and whether the location is in the Northern or Southern Hemisphere. In certain circumstances, upwelling leads to landwards flow at the sea-bed and downwelling leads to offshore flow.

How will these flows affect sediment bedload movement?

There will be landwards or seawards movement of the sediment in response to upwelling and downwelling, respectively.

It takes a period of several hours for wind-driven currents to be established in response to local storms and sometimes there is insufficient

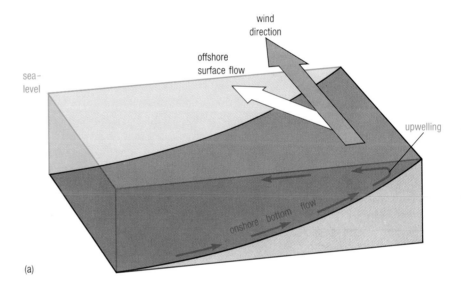

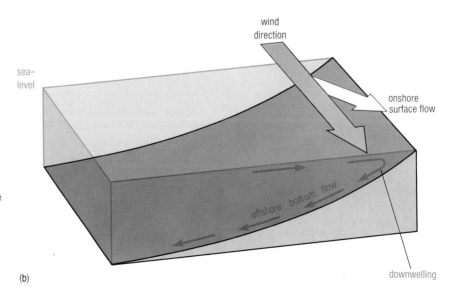

Figure 8.2(a) A southerly shore-parallel wind (in the Northern Hemisphere) resulting in an offshore surface flow at the coast and an onshore flow at the sea-bed with upwelling.

(b) A northerly shore-parallel wind (in the Northern Hemisphere) resulting in an onshore surface flow at the coast and an offshore flow at the sea-bed with downwelling.

time for a current to respond completely to the surface wind stress before the storm has died away. When wind-driven currents are fully established, then in shallow water they are capable of transporting sediment, and near sea-bed speeds of more than $0.25\,\mathrm{m\,s^{-1}}$, rising to around $1\,\mathrm{m\,s^{-1}}$ at a metre above the bed, have been recorded.

Unidirectional currents associated with longshore winds on the Washington-Oregon shelf exceed $0.8\,\mathrm{m\,s^{-1}}$, and lead to the transportation of silt and sand across the shelf towards the shelf break where the sediment may cascade down the continental slope, and be lost into the deep oceans.

8.2.3 TIDAL CURRENTS

In the open ocean the amplitude of the tidal wave is not very great, and the tidal range is correspondingly small—rarely more than 0.5m. When the tidal wave reaches the shallow water of the continental shelf, it behaves like any other wave which moves from deeper water to shallower water.

QUESTION 8.3   What will happen to:

(a)   the tidal wave amplitude?

(b)   the tidal range?

(c)   the tidal current speed?

Thus there is an increased capacity for tidal currents to move sediments on the shelf. On shelves which face the open ocean, the tidal range may increase to 2m or more (Figure 7.2). However, you should now be aware that in more enclosed seas and bays the tidal range may be far greater; e.g. 6.8m in the Wash, 12m in the Bristol Channel and a maximum of 15.4m in the Bay of Fundy. These greatly increased tidal amplitudes result from the phenomenon of resonance (Section 2.4.1).

Consider once again the bath of water illustrated in Figure 1.18(c). If the water is set into oscillatory motion, waves appear which travel down the length of the bath, and are reflected at the end. Similarly, as the tidal wave reaches the head of a bay, or a partially enclosed sea, it is reflected and returned towards the entrance. If the length of the bay, or sea, is just right, the wave is reflected back to the entrance at the same time as the arrival of the next tidal wave. A standing-wave tide is thus produced which is increased in amplitude and has a range that is perhaps several times greater than that of the tide in the open ocean. You may recall that in the case of the Bay of Fundy (Section 2.4.1) the natural resonant period is around 12.5 hours, which is close to that of the semi-diurnal tide and results in the exceptionally high tidal range and strong tidal currents. Resonance is also theoretically possible on open continental shelves, provided that the shelf widths are great enough. It has been found that resonance should occur when the shelf width is about one-quarter of the tidal wavelength. Similarly, resonance will occur for shelf widths 3/4, 5/4, and so forth, of the tidal wavelength. In a relatively shallow sea of about 100m depth, the tidal wavelength for M_2, the principal lunar semi-diurnal tide, is about 1400km.

Although the widths of present-day continental shelves are, in general, less than the one-quarter of the tidal wavelength necessary for resonance, there is an approximately linear relationship between the shelf width and the nearshore tidal range. The wider the shelf, the more closely the conditions approach those required for resonance to occur.

QUESTION 8.4 With the aid of an ocean floor map, or an atlas that shows the widths of the continental shelves, compare the widths of the South American shelf and the African shelf, bordering the South Atlantic. Is there any relationship between shelf width and the coastal tidal ranges shown in Figure 7.2?

The importance of increased tidal ranges is that the tidal current speeds, and the potential for sediment reworking, are increased as well. Where tidal ranges are high, the continental shelves are dominated by tidal current processes. For example, the mean spring near-surface tidal current speeds round the British Isles exceed $1.5\,\mathrm{m\,s^{-1}}$ in places. Close to the sea-bed, these currents are strong enough to rework significant quantities of sand, or even gravel and cohesive mud. In the geological past, there were periods when the sea inundated what are now continental margins. The continental shelves produced were much broader, so high tidal ranges and vigorous tidal currents may have been widespread factors affecting shelf sediments.

It is also important to recall that tidal currents are not steady, unidirectional currents. They accelerate towards and decelerate from a maximum speed twice within each tidal cycle, and they change direction with the flood and ebb tides. As the velocity profile readjusts gradually to changes in the state of the tide, the current speed at the bed lags behind, or is in advance of, that at the surface. Consequently, the actual shear stress at the bed beneath a tidal current is unlikely to be the same as that predicted from observations of surface current speed.

Do you recall how actual shear stress at the bed may differ from the estimated shear stress, and how this may affect predictions about sediment transport?

The value of shear stress is likely to be underestimated for accelerating currents and overestimated for decelerating currents (see Section 4.1.6), leading to underestimates and overestimates, respectively, of the maximum grain size in transport, and the rate of sediment transport.

Much of what we know about sediment transport on tidal shelves is based on research carried out on the continental shelves and epeiric seas around the British Isles, where it is usual for either the peak ebb tidal current to exceed the peak flood tidal current, or vice versa (see Figure 4.10(a) and Section 4.3.1). This situation would be expected on any shelf where there are strong tidal currents. Around the British Isles the situation arises through the interaction between the larger semi-diurnal constituent (M_2) and the smaller quarter-diurnal constituent (M_4). Where the peak currents of the M_2 and M_4 constituents are in phase, the tidal current is strengthened by addition, or weakened by subtraction (Figure 8.3). A similar effect would also occur on shelves where a small diurnal constituent interacted with a larger semi-diurnal constituent. There is no significant net movement of water in the direction of the stronger current.

Thinking back to Section 4.3.1, can you see why not?

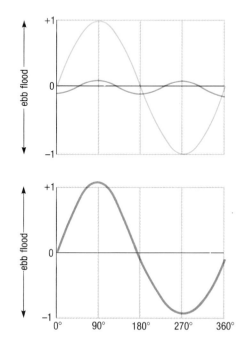

Figure 8.3 The combination of the semi-diurnal, M_2, and quarter-diurnal, M_4, tidal constituents. When the semi-diurnal, M_2, and quarter-diurnal, M_4, tidal constituents are in phase, the flood tide is strengthened and the ebb tide is weakened.

Because the stronger flow is of a shorter duration than the weaker flow. However, there *is* net movement of sediment because not all the sediment moved by the stronger flow will be returned by the weaker flow; the coarsest fraction will be left behind. To give you some idea of the significance of the difference between the flows, the peak ebb and flood mean spring currents may vary by as much as $0.6 \, \text{m s}^{-1}$. But even a difference as small as $0.05 \, \text{m s}^{-1}$ ($5 \, \text{cm s}^{-1}$) is sufficient to determine a definite direction of long-term net sediment transport.

QUESTION 8.5   Can you recall why this should be?

The answer to Question 8.5 is well illustrated by Figure 4.10 and the answer to Question 4.10(b). Maximum transport is seen to occur within about 1.5 hours of the peak ebb and flood currents, and the routes along which net sediment movement occurs are called **sediment transport paths**. These paths are usually several hundred kilometres long and up to a couple of hundred kilometres wide.

Although the directions of sediment transport paths were deduced from the direction of the stronger tidal current, they have been verified using several other complementary lines of evidence, including the migration and asymmetry of observed bed forms (Section 4.5.1), and the systematic decrease in bed grain size where the sediment transport path is

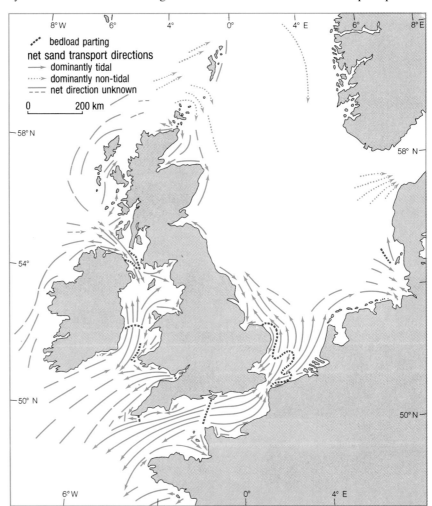

Figure 8.4   The net sediment transport paths on the continental shelf around the British Isles based on a compilation of the available data. Note that the arrows show the direction of net sediment movement, but their length is not proportional to the length of any given transport path.

associated with decreasing peak current strengths; they have also been determined experimentally using the dispersal of artificially introduced mineral grains or fluorescent tracers. Figure 8.4 shows the net sediment transport paths round the British Isles, based on a compilation of the available data.

From Figure 8.4 you can see that the sediment transport paths originate at zones of divergence known as **bedload partings**. These tend to be located in relatively narrow straits where the tidal currents are particularly strong due to horizontal pressure gradients caused by differences in sea-level at either end of the straits (Section 2.4.1). One notable exception occurs at the Straits of Dover where a bedload convergence occurs as sediment transport paths meet. It is probable that the positions of both partings and convergences are determined by the sum of the M_2 and M_4 constituents. Around the British Isles generally, sediment is moved away from a bedload parting towards the open ocean or sea in one direction, and towards the more enclosed sea in the other direction; e.g. outwards towards the Atlantic Ocean and North Sea, or inwards towards the St. George's Channel, Irish Sea and English Channel. Occasionally, transport occurs inwards towards estuaries and embayments, e.g. the Solway Firth and Liverpool Bay.

As you might expect, the rate of sediment transport along a sediment transport path is not constant with time. It increases steadily from the smallest neap tide to the largest spring tide. The actual value of the spring tide will also vary from one year to another, and even on the scale of hundreds or thousands of years.

QUESTION 8.6   Figure 8.5 is a numerical simulation of the magnitude and direction of the maximum bed shear stresses experienced during the twice-daily tidal cycle on the shelf around the British Isles.

(a)   How well do the bed-stress vectors agree with the positions of the bedload partings and sediment transport paths shown in Figure 8.4?

(b)   Calculate the shear velocity from the bed shear stress vector with the greatest magnitude shown in Figure 8.5. What is the size of the coarsest material that might be moved as bedload on the shelf, as the result of tidal current action? (You should assume that all the bed shear stress shown in Figure 8.5 is available for sediment transport. Take the density of seawater to be 1000kg m^{-3}.)

Even if, because of form drag (Section 4.1.6), only half the bed shear stress is available for sediment movement, coarse sand and fine gravel may still be transported in some areas of the shelf.

Most movement of sediment on tide-dominated shelves occurs when the tidal currents are enhanced by wave action (Section 5.2.5).

Do you recall why this happens?

Except in the littoral zone, waves cause very little progressive sediment movement. However, the orbital velocity of a wave is often sufficient to lift into suspension grains which are then transported some distance by currents that are too weak to move the grains unaided, before they are re-deposited. Waves may also lift cohesive muds into suspension, even though tidal currents alone may not shift them.

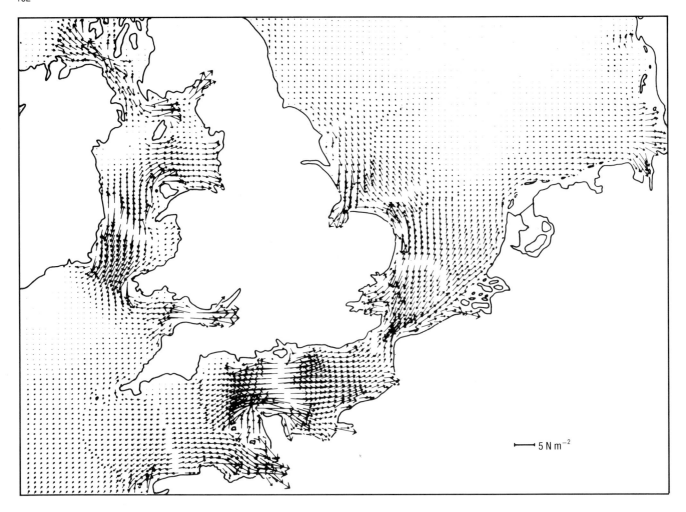

Figure 8.5 A numerical simulation of the magnitude and direction of the maximum bed shear stresses (in Nm^{-2}) experienced during the twice-daily tidal cycle on the continental shelf around the southern British Isles. The length of the arrows is proportional to the maximum bottom stress (see scale bar on Figure).

8.2.4 STORM SURGES

The nature and occurrence of storm surges has already been described extensively in Section 2.4.2. Their importance for sediment movement is that they can cause a local and temporary increase in the tidal range, leading to abnormally high or low tides and increased current speeds. The seawards-returning storm-surge ebb currents may also result in the erosion of nearshore sediments, and their re-deposition in offshore areas.

8.2.5 WAVE ACTION

Because of the enhancement effect described in Section 8.2.3, wave action is important for sediment movement on those shelves where the tidal range is low and tidal currents are weak. During fair weather conditions, sediment is disturbed by wave action at water depths of around 10 to 20m at low tide. This depth defines the seawards edge of the littoral zone (Chapter 5). On shelves where storms are frequent, very occasionally sediments may be affected by wave action at depths as great as 200m. For example, it has been estimated that wave action on the Washington shelf affects sediments at a depth of 75m (the mid-shelf area) for approximately 53 days per year. Sediments at the shelf break, a depth of about 167m, are reworked by waves as often as five days per year. Such storm events are capable of influencing shelf sedimentation processes out of all proportion to their infrequent occurrences.

8.3   BED FORMS ON THE CONTINENTAL SHELF

Bed forms are some of the most important indicators of the speed and
direction of sediment transport on the tide-dominated continental shelf.
The current flow speeds decrease along sediment transport paths,
consequently a series of bed forms develops along a sediment transport
path which reflects the direction of the waning current (Figure 8.6 and
Section 4.5.2).

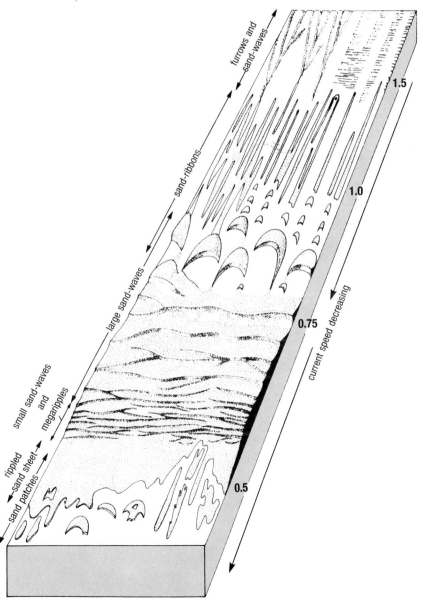

Figure 8.6   Block diagram to show the bed forms
developed along a sediment transport path with
decreasing current speed (given in ms⁻¹). Note that
the distance covered by the diagram spans hundreds
of kilometres.

At near-surface peak current speeds of more than 1.5 down to $1.0\,\mathrm{m\,s^{-1}}$, the bed forms are linear and approximately parallel to the direction of current flow (scour hollows, furrows and sand-ribbons). At lower current speeds, the bed forms are transverse to the direction of current flow (sand-waves, megaripples and current ripples). As you have already seen (Section 8.2.3), the asymmetry of transverse bed forms also provides evidence for the direction of sediment movement.

8.3.1 SCOUR HOLLOWS, FURROWS AND SAND-RIBBONS

Where the near-surface peak current speeds exceed $1.5\,\mathrm{m\,s^{-1}}$, elongate **scour hollows** may develop on the sea-bed (these are not shown on Figure 8.6). In the North Channel of the Irish Sea, scour hollows up to 28 m deep have been found. Others have been found at the entrance to San Francisco Harbour, and within the Hayasui Strait between the islands of Kyushu and Shikoku in Japan. The features of the Irish Sea dwindle into insignificance compared with those in the Hayasui Strait (see Figure 8.7).

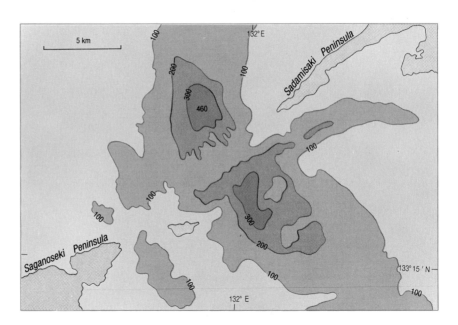

Figure 8.7   Scour hollows more than 260 m and 160 m deep (northern and southern hollows, respectively) below the surrounding sea-floor in the Hayasui Strait, Japan. Depths below the sea-bed are given in metres.

Where the near-surface peak current speeds are around $1.5\,\mathrm{m\,s^{-1}}$, sand is winnowed away completely, either exposing the bed rock, or leaving a residue of gravel worked into linear, bifurcating **furrows**. These furrows extend down-current for several kilometres, are up to 1 m deep and more than 10 m wide (Figure 8.8). As linear **sand-ribbons** extend down-current from the furrows, the furrows are probably passageways for the removal of sand. Sand-ribbons vary greatly in size, reaching maximum dimensions of around 15 km length, 200 m width and 1 m thickness. As the current speed decreases downpath, the sand-ribbons pass into ribbon-like trains of crescentic sand-waves.

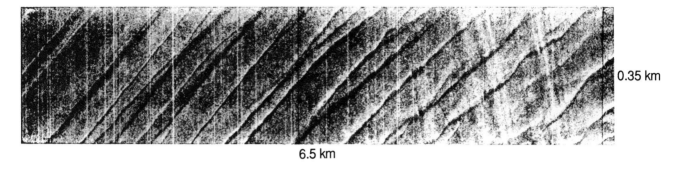

0.35 km

6.5 km

Figure 8.8 Sonograph showing longitudinal furrows in gravels in the English Channel.

8.3.2 SAND-WAVES AND OTHER TRANSVERSE BED FORMS

At current speeds of less than about $0.75\,\mathrm{m\,s^{-1}}$, **sand-waves** occur extensively. For example, they cover an area of $15000\,\mathrm{km^2}$ off the Netherlands coast. Sand-waves may reach 18m high and have wavelengths of nearly a kilometre, although most are much smaller (e.g. see Figure 8.9). Their amplitude is enhanced by the oscillatory nature of tidal flow. Once formed, sand-waves may migrate along a sediment transport path at rates of between 10 and 150m per year. The largest sand-waves constitute a considerable hazard to shipping where they occur in shallow water because they may stand well above the level of the charted sea-bed. Many shelf seas were surveyed hydrographically when vessels generally had draughts of less than 17m, and the charts made then are still in use. As vessels have increased progressively in tonnage and draught, they run an increasing risk of grounding on large, uncharted sand-waves.

Large sand-waves frequently have ripples or megaripples superimposed, especially on the gentle, upstream slope (Figure 8.9). These smaller bed forms develop even though the current speed and overall bed shear stress are greater than those at which you would expect ripples or megaripples to form on a flat sea-bed at the water depths of the continental shelf.

QUESTION 8.8   Bearing in mind that the sea-bed is *not* flat, can you suggest why these bed forms may develop at higher current speeds than expected?

Figure 8.9 Sonograph of small sand-waves situated on the flank of a large sand-wave.

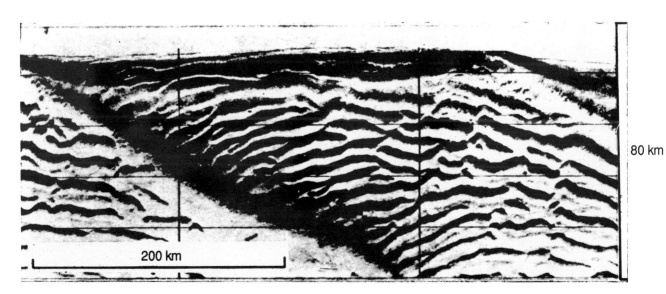

80 km

200 km

As current speeds decrease, the larger sand-waves pass into smaller sand-waves and megaripples, both of which frequently have current-formed ripples on their backs. Contrary to what you might expect intuitively, and from the discussion of larger sand-waves, it is most probable that the smaller current ripples form first, and may even initiate the formation of the megaripples. You may recall that current ripples develop only where the grain roughness does not totally destroy the viscous sublayer (Section 4.5.2). Once formed, however, current ripples increase bed roughness, so that turbulence extends right down to the sea-bed, generating the megaripples without erasing the current-rippled surface.

8.3.3 SAND-BANKS

Wherever there is abundant sand on the sea-bed, and tidal current flows exceed about $0.5\,\mathrm{m\,s^{-1}}$, **sand-banks** can be expected to develop. These are large topographical features on many continental shelves and are quite distinct from the swash bars that develop in the littoral zone (Section 5.1.3), which are often wrongly called 'sand-banks'. True sand-banks are submarine features which may extend up to 80km in length, 3km in width and tens of metres in height. Like the large sand-waves, they are a hazard to shipping, e.g. around the British Isles, the Goodwin Sands, north of the Straits of Dover, are particularly notorious.

8.4 SHELF PROCESSES AND SEA-BED RESOURCES

In this Section we are concerned only with those resources which occur within the mobile, or unconsolidated, sediments of the continental shelf, and which are affected directly by shelf or shelf break processes. The locations of major identified sea-bed resources are widespread (Figure 8.10). They are grouped into three main types: aggregates, placer deposits and bedded phosphorites. **Aggregates** comprise mainly sand and gravel which are used by the construction industry, but also include shell deposits. **Placer deposits** are mostly heavy metallic minerals, and **phosphorites** are mainly related to oceanic upwelling close to the continental shelf break. Of these, aggregates and tin placer deposits have been the only shelf resources mined on a large scale (Table 8.1). Despite their relatively small economic contribution, sea-bed resources have generated considerable interest, not simply because of their potential value, but also because of the problems concerning their legal ownership and extraction.

Table 8.1 The production of some shelf resources compared to the world's production of these resources (based on data available in 1987).

Resource	Sea-bed production ($\times 10^3$tonnes)	World production ($\times 10^3$tonnes)
Sand and gravel	112300	7620480
Shells (for $CaCO_3$)	16667	1666667
Placer deposits (mainly tin ore)	28	201

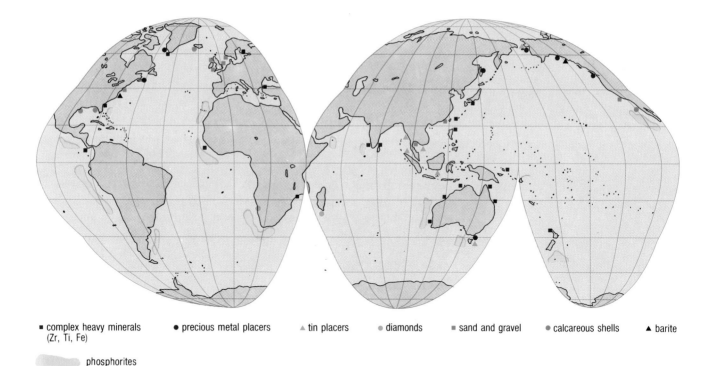

■ complex heavy minerals
(Zr, Ti, Fe)　　● precious metal placers　　▲ tin placers　　● diamonds　　■ sand and gravel　　● calcareous shells　　▲ barite

phosphorites

Figure 8.10　The approximate locations of identified aggregates, placer deposits and phosphorites.

8.4.1　AGGREGATES

Shelf deposits represent important aggregate resources as they can be recovered without direct damage to the countryside although, inevitably, there is some detrimental effect on the marine environment itself. For example, where the sands and gravels contain a high proportion of fine sediment, disturbance of the sediment may cloud the water, reduce light penetration and smother bottom-dwelling organisms. This is a serious problem in those offshore regions where reefs flourish, e.g. off southern Florida and round many Caribbean islands.

QUESTION 8.9　Refer back to Section 5.2.4. Can you suggest why dredging for sand and gravel in the littoral zone might prove harmful?

It is important to restrict dredging to those areas which are beyond the littoral zone, or to where the natural sediment input is sufficiently high to replace that lost by dredging. Japan is the largest producer of offshore sand and gravel (Table 8.2), followed by the UK, Denmark and the USA. Normally, sands and gravels can be recovered economically only from water depths less than about 35 m because of the costs of the heavy equipment necessary to dredge in deeper water. However, Japan has developed technologies which permit exploitation in water depths up to 50 m. Around the UK shelf, the annual extraction of sands and gravels has exceeded 13×10^6 tonnes since 1968.

Where would you expect the main sites of offshore exploitation to occur?

Figures 8.4 and 8.5 suggest that the most likely sites are the zones of bedload parting, and the bedload convergence in the Straits of Dover, where the maximum bottom stress values in opposite directions are

Table 8.2　Production of sand and gravel from the sea-bed in 1978.

Country	Production ($\times 10^3$ tonnes)
Japan	84 700
UK	15 000
Denmark	4 600
USA	4 000
France	2 800
The Netherlands	2 430
Belgium	1 140
Iceland	600

highest. In fact there are six main dredging areas which occur off the east and south coast of the UK, in the Thames Estuary, the Bristol Channel, the Humber Estuary and Liverpool Bay. However, many potential locations for dredging cannot be exploited because of the risk to the dynamic equilibrium of the coastal zone and also to fisheries. In addition, the presence of underwater cables and pipelines has forced other reserves to remain untapped, especially in the North Sea.

Shell material is a valuable source of calcium carbonate in various parts of the world where there is a deficiency of terrestrial limestone. As shell material is being replaced continually by lime-secreting organisms, it is, effectively, a constantly renewable resource. Exploitation occurs on a large scale from the Great Bahama Bank (for use in the USA), in the Gulf of Mexico, and round Iceland where shell beds and algal debris are dredged (to the detriment of reefs or other bottom-dwelling organisms), for use in cement manufacture and for agricultural purposes.

8.4.2   PLACER DEPOSITS

Placer deposits consist of dense, resistant and often economically valuable minerals which have been weathered from terrigenous rocks, transported to the sea and concentrated in marine sediments by wave or current reworking. Many of these minerals occurred originally as placer deposits in river sediments, or were concentrated in beach sands, when the sea-level was much lower than it is now. As they have higher densities than that of sand, they are concentrated within the coarse sands and gravels after the finer sediment has been winnowed away.

The tin-bearing mineral, cassiterite (SnO_2), has been the most important offshore placer deposit for nearly a century. Tin dredging has taken place off Thailand since 1907 and, as the onshore deposits of tin in south-east Asia approach exhaustion, exploration is now concentrated on offshore deposits. Some 2000 years of tin mining in Cornwall has led to the fluvial transportation of tin-bearing sands and gravels to the coast to form artificial cassiterite placer deposits. Attempts to recover the ore by dredging have been thwarted by the Atlantic storm waves which, although helping to concentrate the cassiterite, have also prevented the dredgers from working during the winter months, rendering the operation uneconomic.

Figure 8.10 shows the locations of other known placer deposits, of which relatively few have been worked. Between 1961 and 1971, diamonds were dredged at Hottentot Bay, near Orange River, Namibia, but harsh weather conditions and severe storms led to loss of the dredging apparatus.

Although the exploitation of sea-bed placer deposits is limited, heavy minerals of commercial value have been extracted from beach sands. For example, the mineral monazite, $(Th,La)PO_4$, as a source of thorium, has been recovered from beach sands in eastern and western Australia. It is also widespread along the west coast of the USA, and the coasts of India and Brazil. Similarly, ilmenite ($FeTiO_3$) and rutile (TiO_3), important sources of titanium, are extracted from Australian and Californian beach sands.

8.4.3 PHOSPHORITES

Marine phosphorites are widespread on continental shelves close to the shelf break and they contain economic concentrations of calcium phosphate minerals used for example in fertilizers. Most deposits are believed to have formed in regions of upwelling close to the continental margin. During upwelling, subsurface waters rich in nutrients, including phosphorus, are brought to the surface. This enhances productivity in the surface waters and results in increased extraction of phosphorus from seawater by marine organisms to form biological tissue. When the organisms die, their soft parts decay and the phosphorus is returned to the seawater. However, since the process of decay occurs slowly as the organisms sink through the water column, seawater is often enriched in phosphorus down to the depths of the abyssal plains. As a result, the water trapped between the sediment grains on the continental slope and shelf break is also enriched in phosphorus which replaces the carbon in the calcium carbonate of shell materials to form calcium phosphate.

Would you expect economic deposits of phosphorites to occur wherever there is ocean margin upwelling?

No, because there must be abundant accumulation of shell material, and this will only take place where the input of terrigenous material is sufficiently low to prevent the dilution of the organic calcareous debris.

8.5 SUMMARY OF CHAPTER 8

1   Continental shelves are submerged continental margins whereas epeiric seas (or epicontinental platforms) are flooded areas within a continent. However, the same processes of waves and current action affect both.

2   The nature of shelf sediments is largely influenced by climate, with sands and gravels dominant at higher latitudes and muddy sediments dominant at lower latitudes. Carbonate sediments accumulate where the supply of terrigenous sediments is negligible.

3   The main processes affecting sediment movement in shelf seas are tidal currents and storm waves, but ocean currents (when they migrate onto the shelf), local wind-driven currents and storm surges may cause sediment movement.

4   Tidal currents are significant where the tidal range is high. The amplitude of the tidal range may be increased by resonance, which may occur in coastal embayments, partially enclosed seas and wide continental shelves. Inequalities in the strength of the ebb and flood peak tidal currents lead to net sediment movement along sediment transport paths which originate at zones of bedload partings.

5   A series of bed forms develops along a sediment transport path which reflects the direction of the waning current. Scour hollows develop at the highest speeds. With decreasing speed a sequence of linear furrows and sand-ribbons are formed, passing into transverse sand-waves, megaripples and ripples. The asymmetry of transverse bed forms is an indication of the direction of sediment movement.

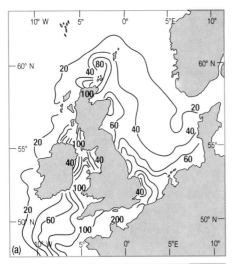

(a)

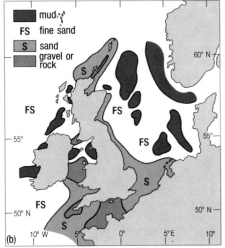

mud

FS fine sand

S sand

gravel or rock

(b)

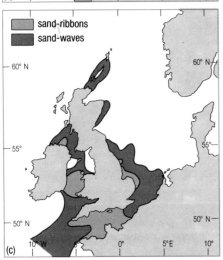

sand-ribbons

sand-waves

(c)

Figure 8.11(a) Distributions of the mean near-surface tidal currents around the British Isles (values in ms^{-1}).

(b) Distributions of shelf sediments around the British Isles.

(c) Distributions of sand-ribbons and sand-waves on the shelf around the British Isles.

6   The main sea-bed resources resulting from shelf processes are aggregates (sands, gravel and shells), placer deposits and phosphorites. Sands and gravel are exploited on a large scale in water depths up to 50m, but their extraction presents potential hazards to the marine environment including the dynamic equilibrium of the littoral zone. Shell material is an important source of $CaCO_3$ for countries deficient in limestone. The most important marine placer deposit is cassiterite (tin ore) which has been dredged off south-east Asia for nearly a century. Most phosphorites have originated along shelves where oceanic upwelling brings nutrient-rich subsurface (~100–200m) waters to the surface, enhancing productivity.

Now try the following questions to consolidate your understanding of this Chapter.

QUESTION 8.10   Figure 8.11 consists of simplified maps of the seas around the British Isles showing the distributions of the mean near-surface spring tidal currents, sediment types and sand-waves and sand-ribbons, respectively.

(a)   Comment on the relationship between the average tidal current speed and the sediment types.

(b)   Comment on the relationship between the average tidal current speeds and the distributions of sand-waves and sand-ribbons.

(c)   To what extent are the distributions of sediments and bed forms consistent with the directions of the sediment transport paths shown in Figure 8.4?

(d)   No furrows are shown in Figure 8.11(c), but whereabouts might you expect to find them?

QUESTION 8.11   Explain how you can deduce the tidal current directions responsible for the scour hollows shown in Figure 8.7.

QUESTION 8.12   To what extent is it justifiable to consider marine placer deposits as both residual and relict sediments?

SUGGESTED FURTHER READING

ALLEN, J. R. L. (1985) *Principles of Physical Oceanography*, Allen and Unwin. A moderately mathematical introduction to the physics of fluid flow and sediment movement, illustrated with simple laboratory experiments.

DYER, K. R. (1986) *Coastal and Estuarine Sediment Dynamics*, John Wiley and Sons. An advanced textbook which integrates marine sedimentology with experimental results and theory from oceanography. The approach is quantitative and moderately mathematical, and the problems of applying theories to the real marine environment are brought out.

GROEN, P. (1967) *The Waters of the Sea*, Van Nostrand. Useful background material, including a good general coverage of waves and tides.

KINSMAN, B. (1965) *Wind Waves: Their Generation and Propagation on the Ocean Surface*, Prentice-Hall. A more detailed survey of the dynamics of wind waves, with a more mathematical treatment than that given in this Volume.

KNAUSS, J. A. (1978) *Introduction to Physical Oceanography*, Prentice-Hall. Provides a broad and clear qualitative picture of most aspects of the subject, including waves and tides. The descriptive aspects are complemented by relevant equations and a mathematical treatment where appropriate.

LEEDER, M. R. (1982) *Sedimentology: Process and Product*, Allen and Unwin. Provides a more geological approach to physical sedimentology and covers a wide range of topics, including the origins of sediment grains, continental and deep-sea environments and sediment diagenesis, in addition to the physics of sediment movement.

MINISTRY OF AGRICULTURE, FISHERIES AND FOOD (MAFF) (1981) *Atlas of the Seas around the British Isles*, HMSO. Complete set of maps and descriptions of waves, tides, tidal currents, sediment distribution and transportation, and bed forms, as well as temperature and salinity structures and a host of other topics including fisheries, pollution, marine transport and communications.

PETHICK, J. (1984) *An Introduction to Coastal Geomorphology*, Edward Arnold. Do not be misled by the use of 'geomorphology' in this title. This is a simple introduction to waves, tides, coastal sediments, the littoral zone and estuaries which is easily accessible to the mathematically less confident reader.

PUGH, D. T. (1987) *Tides, Surges and Mean Sea-Level*, Wiley. A comprehensive explanation of tides, surges and sea-level changes and their effect upon shallow water and coastal environments. The book is written for marine and coastal engineers, hydrographers, sedimentologists and similar professions, and has a strong practical bias.

ROBINSON, I. S. (1985) *Satellite Oceanography*, Ellis Horwood. This book is a 'state of the art' review and appraisal of the value of satellite-based techniques to our future knowledge and understanding of oceanography.

CHAPTER 1

Question 1.1   Five seconds. The frequency is $0.2s^{-1}$, i.e. during one second '0.2 of a wave' passes a fixed point. To find out how long it takes for the whole wave to pass (the period), we need to divide 1 by $0.2s^{-1}$. So:

$$T = 1/0.2 = 5s.$$

Question 1.2   The less steep of the two. If steepness $= H/L$, then $H =$ steepness $\times$ wavelength. Therefore, because H is the same for both waves, the less steep wave will have the greater wavelength, and hence travel faster.

Question 1.3   It decreases. The '40-knot' spectrum contains much more energy than either of the other two spectra. Most of this energy is related to the low frequency waves that a 40-knot wind would generate.

Question 1.4   Sixteen waves in 64 seconds = a period of 64/16 seconds = 4s. The frequency is thus the reciprocal of 4s, i.e. $0.25s^{-1}$.

Question 1.5   First, convert frequency to period. Period $= 1/0.05 = 20s$.

(a) $-a$ (a trough at P.)

(b) $+a$ (a peak at P.)

(c) 0 (η changes from $+a$ to $-a$ in 10 seconds, so five seconds after a peak $(+a)$ the displacement is zero.)

(d) 0

(e) $+a$

(f) 0

Note for (d), (e) and (f):
The distance between P and Q is half a wavelength. Note that if the displacement at P is zero and is diminishing, then the displacement at Q is zero and is increasing (and vice versa).

Question 1.6   If $k = 2\pi/L$, and $\sigma = 2\pi/T$,
then $L = 2\pi/k$, and $T = 2\pi/\sigma$.
Substituting into $c = L/T$, we have:

$$c = \frac{2\pi/k}{2\pi/\sigma} = \frac{1/k}{1/\sigma} = \frac{\sigma}{k}$$

Question 1.7(a)   If d is greater than $0.5L$, then $2d$ is greater than L, and the expression $2\pi d/L$ becomes greater than π. The tanh of numbers greater than π approximates to 1. So tanh $(2\pi d/L) \approx 1$ and equation 1.3 approximates to:

$$c = \sqrt{\frac{gL}{2\pi}}$$

(b) If d/L is very small, then $2\pi d/L$ is also very small, and hence tanh $(2\pi d/L)$ approximates to $2\pi d/L$.

So equation 1.3 becomes:

$$c = \sqrt{\frac{gL2\pi d}{2\pi L}}$$
$$= \sqrt{gd}$$

Question 1.8 From equation 1.1, $c = L/T$.
From equation 1.4, $c = \sqrt{gL/2\pi}$
So $\sqrt{gL/2\pi} = L/T$, and $gL/2\pi = L^2/T^2$
From which $L/T^2 = g/2\pi$ and $L = gT^2/2\pi$.

Question 1.9 If, in equation 1.6, $g = 9.8\,\text{ms}^{-2}$, and $\pi = 3.14$, then $g/2\pi = 1.56\,\text{ms}^{-2}$.
Substituting this value into $L = gT^2/2\pi$, we get $L = 1.56\,\text{ms}^{-2} \times T^2$, and because T is in seconds, then $L = 1.56T^2$ (m).

Question 1.10(a) $31.2\,\text{ms}^{-1}$. You may have done this the hard way by $L = 1.56 \times 20 \times 20 = 624$, followed by $c = 624/20 = 31.2\,\text{ms}^{-1}$. Better still, you may have derived the formula $c = 1.56T$ from equations 1.1 and 1.6, and done the sum in one step.

(b) $22.1\,\text{ms}^{-1}$. From equation 1.4,
$$c = \sqrt{1.56 \times 312}$$
$$= \sqrt{486.7}$$
$$= 22.1\ \text{ms}^{-1}$$

(c) $10.8\,\text{ms}^{-1}$ in *both* cases.
Remember, if the depth is less than 1/20 of the wavelength, all waves will travel at the same depth-determined speed in shallow water, i.e. the depth is the only controlling factor. So, from equation 1.5, we get:
$$c = \sqrt{gd}$$
$$= \sqrt{9.8 \times 12}$$
$$= \sqrt{117.6}$$
$$= 10.8\,\text{ms}^{-1}$$

Question 1.11 No. It would be quadrupled, because the energy of a wave varies with the square of the wave height (equation 1.10), and hence with the square of the wave amplitude.

Question 1.12(a) If amplitude is 1.3m, then wave height is 2.6m. The values of the constants g and ρ, and also the above value for wave height, can be plugged into equation 1.10, giving:
$$E = 1/8 \times 1.03 \times 10^3 \times 9.8 \times 2.6^2$$
$$= 8.5 \times 10^3 \ \text{Jm}^{-2}$$

(b) The wave power per unit length is the product of the wave energy per unit area and the group speed. We know the wave energy from (a) above, and can calculate the group speed from the height and steepness as follows:

steepness (0.04) = height (2.6m)/wavelength.

So wavelength $= 2.6/0.04 = 65\text{m}$.

From equation 1.4, wave speed, $c = \sqrt{gL/2\pi}$

So $c = \sqrt{1.56L}$
   $= \sqrt{1.56 \times 65}$
   $= 10.07\,\text{ms}^{-1}$.

From which group speed $= 10.07/2$
   $= 5.03\,\text{ms}^{-1}$.

So wave power $= 8.5 \times 10^3\,\text{Jm}^{-2} \times 5.03\,\text{ms}^{-1}$.
   $= 42.7\,\text{kWm}^{-1}$.

Question 1.13   Spectrum (a) on Figure 1.12 shows the wave energy distributed amongst a wide range of frequencies. The peak is rather poorly defined, and hence must represent the storm-generating area. Spectrum (b), on the other hand, has a much narrower range of frequencies, and a clearly defined peak. It thus represents the regular swell waves at a point well away from the storm.

Question 1.14   The wave refraction diagram (Figure 1.14(b)) illustrates how the offshore Hudson Canyon is effective in defocusing storm waves as they approach the Long Branch coastal section from the east–south-east, and in refracting them onto other beaches. Fishermen can leave their boats on the beach at Long Branch during all seasons of the year, despite its apparent exposure to the full force of Atlantic storms. The wave rays are, if anything, focused as they enter the mouth of the Hudson River, so that the energy of storm waves would certainly not be diminished there, and might even be increased. People leaving their boats in this apparently sheltered region could be courting disaster.

Question 1.15   If the change in the beach slope was sufficient, you might expect to see collapsing breakers, and if it got really steep, surging breakers as well.

Question 1.16   You have no information on the length of the ship, but if you calculate the wavelength corresponding to a period of 30s, using the expression you derived in Question 1.9 you get $L = 1.56T^2 = 1.56 \times 900 = 1\,404\,\text{m}$. You might conclude that the sailor is trying to tell you his ship was over 700m long (nearly half a mile). The longest super-tankers are only about 320m. However, return to the main text and read on. . . .

Question 1.17   *Exeter* was 175m in length, so if the story were true, the wavelengths concerned were 350m. The ship was travelling at $11.8\,\text{ms}^{-1}$, so in 30 seconds it would have travelled $30 \times 11.8\,\text{m} = 354\,\text{m}$. In 30 seconds, an overtaking wave would have travelled one wavelength plus the distance the ship had travelled, i.e. $354 + 350 = 704\,\text{m}$, and the wave speed would be $704/30 = 23.5\,\text{ms}^{-1}$.

From equation 1.4, we can find the wave speed corresponding to a wavelength of 350m, i.e.

$$c = \sqrt{gL/2\pi}$$
$$= \sqrt{1.56 \times 350} = 23.4\,\text{ms}^{-1}$$

which means the sailor's tale is at least consistent with simple wave theory. Full marks if you suspected something of this sort while attempting Question 1.16.

Question 1.18 Because the wavelength is very long compared with an ocean depth of 5 500m over the abyssal plains, the tsunami must be treated as a shallow water wave (equation 1.5):

$$c = \sqrt{9.8 \times 5\ 500} = 232\,\mathrm{ms}^{-1}.$$

Question 1.19 Using equation 1.17, $l = 90\mathrm{m}$, $d = 10\mathrm{m}$.
So $T = 4 \times 90/\sqrt{9.8 \times 10}$
$\qquad = 36.36$ seconds.
Because the resonant period of the harbour is close to 36s, waves of period 18s (half of 36s) would set up a standing wave in the harbour.

Question 1.20 Table 1.1 gives the relationships between wind force and significant wave heights.

(a)(i) $\pm$ 8cm, because $H_{1/3}$ is about 0.15m, which is below 8m.
(ii) $\pm$ 0.3m, because $H_{1/3}$ is below 3m.
(b)(i) $\pm$ 8cm, because $H_{1/3}$ is about 1.5m, which is below 8m.
(ii) $\pm$ 0.3m, because $H_{1/3}$ is below 3m.
(c)(i) $\pm$ 12cm (1cm for each m of $H_{1/3}$, which is about 12m).
(ii) $\pm$ 1.2m (10% of $H_{1/3}$).

Note that the extent of the cloud cover is irrelevant, because water vapour is transparent to radar pulses, and will not affect the signals.

Question 1.21(a) $L = gT^2/2\pi$ (equation 1.6).
Substituting for L in equation 1.1, we get:

$$c = L/T = (gT^2/2\pi)/T$$
$$= gT/2\pi = 1.56T$$

So wave speed, $c\ \ = 1.56 \times 10 = 15.6\,\mathrm{ms}^{-1}$, and
$\qquad$ group speed, $c_g = c/2 = 7.8\,\mathrm{ms}^{-1}$.
(b) From your answer to Question 1.9:
wavelength, $L = 1.56 \times T^2 = 156\mathrm{m}$
So steepness, $H/L = 1/156 = 0.0064$

(c) Wave energy is $1/8\ (\rho g H^2)$ (equation 1.10)
$\qquad = 1/8 \times 1.03 \times 10^3 \times 9.8 \times 1$
$\qquad = 1.26 \times 10^3\ \mathrm{Jm}^{-2}$

From Section 1.4.1, wave power is the product of the wave energy per unit area and the group speed:

$\qquad = 1.26 \times 10^3 \times 7.8$
$\qquad = 9.8\,\mathrm{kWm}^{-1}$

(d) The wave power per metre of wave crest will be exactly the same as the answer in part (c) above. You may have found this out the hard way by calculating the wave height and wave energy in water 2.5m deep. Refer back to equation 1.13 and remember that wave power $= Ec_g$.

Question 1.22(a) Because swell waves and short waves have little interaction (Section 1.4.2), they can be treated separately. From equation 1.4, the short waves will have a speed of:

$$\sqrt{1.56 \times 6} = 3.06\,\mathrm{ms}^{-1}.$$

Hence, group speed is 1.53ms^{-1}, and these waves will not propagate against a current of 3 knots (1.54ms^{-1}), because this current exceeds half the group speed of the waves. In the narrow inlet, these smaller waves will increase in height until they become unstable and break. Because $c = 1.56T$ (see answer to Question 1.21(a)), the swell waves will travel at a speed of 34.3ms^{-1}. Therefore $c_g = 17.15 \text{ms}^{-1}$, and because the counter-current is less than half the group speed, these waves will propagate against the current, but will show an increase in height and a reduction of wavelength.

The wavelength of the swell waves can be found from the period ($L = 1.56T^2$) (see answer to Question 1.9), i.e. $L_0 = 1.56 \times 22^2 = 755 \text{m}$. If we now apply equation 1.15, the wavelength in the narrow inlet can be found.

The period of $22\text{s} = L_{\text{current}}/(34.3 - 1.5)$.
So $L_{\text{current}} = 22 \times 32.8 = 722 \text{m}$.

(b) Beyond the inlet, only the swell waves will be encountered. The waves will have lost height and steepness, because they are no longer propagating against a current. The period of 22s and the 'new' wavelength are not consistent with the relationship between T and L derived from a combination of equations 1.1 and 1.4 (i.e. the answer to Question 1.9: $L = 1.56T^2$). Once beyond the influence of the current, the waves will tend to re-establish those relationships between c, L and T which are typical of sinusoidal waves, although the waves emerging from the straits will not have the same set of characteristics as they had before entry.

Question 1.23(a) Wave speed = apparent speed of overtaking waves, plus the ship's speed.

Apparent speed of waves = $146 \text{m}/6.3\text{s} = 23.17 \text{ms}^{-1} = 45$ knots.

Added to ship's speed of 5.14ms^{-1} (10 knots), this gives an actual wave speed of 28.31ms^{-1} (55 knots).

(b) From equation 1.4, $c = \sqrt{gL/2\pi}$
So, $L = 2\pi c^2/g$
     $= 0.64c^2$
i.e. wavelength $= 0.64 \times \text{(wave speed)}^2$
                 $= 0.64 \times 28.31^2$
                 $= 512 \text{m}$.
Wave height = 34m (given).
Steepness $= 34/512 = 0.066$.

(c) $c = 1.56T$ (answer to Question 1.21(a)), so $T = 0.64c$ and, from part (a), wave speed $= 28.31 \text{ms}^{-1}$. So the period $= 0.64 \times 28.31 = 18.1\text{s}$. This answer is some 22% more than the period of 14.8s reported by the *Ramapo*.

(d) A wave of period 14.8s would have a wavelength of $1.56 \times 14.8^2 = 342 \text{m}$. The steepness of such a wave is $34/342 = 0.1$. This is a very steep wave indeed.

(e) The speed of the waves calculated in part (a) was 28.3ms^{-1} (55 knots). This is exactly the same wave speed which was reported by the *Ramapo*, and there is no reason to doubt it. The discrepancy between the period consistent with such a wave speed, and that reported, may reflect the fact that the observed waves were not simple sinusoidal waves. As waves get steeper, so their shape becomes trochoidal rather than sinusoidal, and the simplifying assumptions made in this Chapter no longer apply.

Question 1.24(a)   After a long calm spell, only low waves of long period would be arriving. Such waves would break as surging breakers on a beach of intermediate slope.

(b)   Winds of this force would generate steep waves, which on a beach of intermediate slope would break as spilling breakers.

CHAPTER 2

Question 2.1(a)   The magnitudes of the gravitational and centrifugal forces would be the same as at C and G on Figure 2.2. The resultant tide-producing force would be directed into the Earth (into the plane of the page as you look at Figure 2.2).

(b)   This is the only point where the gravitational force exerted by the Moon on the Earth is exactly equal to and acting in exactly the opposite direction to the centrifugal force. The resultant tide-producing force at that point is zero.

Question 2.2(a)   The waves would be required to travel 40 000km in 24.83 hours, i.e. 1 611 km hr^{-1}, or 447 m s^{-1}.

(b)   We have to use equation 1.5 for waves of this length. If $c = \sqrt{gd}$, then $d = c^2/g$. So, depth required $= 447^2/9.8 = 20\,389$ m, i.e. more than 20 km.

Question 2.3   Nil. Seven days (i.e. one-quarter of 27.2 days) after the scenario shown on Figure 2.5, the Moon will be overhead at the Equator, and there will be no lunar component of diurnal inequality anywhere on the globe.

Question 2.4   At the summer and winter solstices, i.e. June 21/22 and December 21/22. On these dates the Sun is at maximum declination, and is overhead at the Tropics of Cancer and Capricorn respectively.

Question 2.5(a)   14.75 days.

(b)   Neap tides. Spring tides coincide with syzygy, so 14.75 days after that there will be another spring tide, and 7.4 days more will bring the cycle to neap tide.

(c)   3–4 days, i.e. half-way between the spring tide associated with the new Moon and the neap tide which will occur 7.4 days after the new Moon.

(d)   If the simplification were true, Figure 2.6(a) would show a solar eclipse (Moon directly between Sun and Earth) and Figure 2.6(c) would show a lunar eclipse (Earth's shadow on Moon).

Question 2.6(a)   At a latitude of 26°, the linear velocity on the Earth's surface with respect to the Moon is about 400 m s^{-1}. The speed at which the tidal wave can cross an ocean of 4 080m depth is about $200\,\mathrm{m\,s^{-1}}$ (from equation 1.5). Hence the tide could only move at half the linear velocity of the Earth's surface, and there would be a tidal lag of 6 hours 12 minutes, resulting in the tides being 90° out of phase with the predicted position of the equilibrium tide.

(b)   The linear velocity of the Earth's surface at latitude 10° is about $442\,\mathrm{m\,s^{-1}}$—more than twice the speed of the tidal wave.

You may have answered 6 hours 48 minutes, or thereabouts, for the tidal lag. If so, please note that the lag cannot exceed 6 hours 12½ minutes, because once the tidal bulge has lagged a quarter of an Earth's turn out of phase, it is being turned *towards* the tidal-generating force on the other side of the Earth, and hence the tide-generating forces are now pulling the tidal bulge in the opposite direction. The tidal bulges will thus become established at right angles to the predicted positions of the theoretical equilibrium tides.

Question 2.7(a)   About half a lunar hour after low tide (or five-and-a-half lunar hours before high tide).

(b)   About two lunar hours before high tide (or four lunar hours after low tide).

If you had difficulty with (a) and (b), note that if high tide is at '00' on Figure 2.8, it will take two more lunar hours to reach the Firth of Forth, so that area is expecting a high tide in two hours' time. Similarly, the high tide will take five-and-a-half hours to reach the Wash.

(c)   The tidal range of the Wash (over 4m) exceeds that of the Firth of Forth (more than 3m, but less than 4m).

Question 2.8(a)   A high value of F corresponds to a dominant diurnal tidal component (period of 24 hours and 50 minutes). A low value of F indicates the dominance of a semi-diurnal component.

(b)   Yes. Syzygy occurs every 14.75 days.

Question 2.9   Using equation 1.16, with $l = 270$km, and $d = 60$m, we get:
period of basin $= 2 \times 270 \times 10^3/\sqrt{9.8 \times 60}$
$= 22\ 269$ seconds
$= 6.19$ hours
A simple multiple ($\times 2$) of this period, i.e. 12.38 hours, is very close to the semi-diurnal period, i.e. the period of the basin is a harmonic of the semi-diurnal period, and is consistent with the very large tidal range found in the Bay of Fundy.

Question 2.10   If a 10m column of water ($= 1\ 000$cm) corresponds to one atmosphere of pressure (1 000 millibars), then one centimetre of water corresponds to one millibar. A reduction in atmospheric pressure of 50 millibars would therefore result in a rise in sea-level of 50cm, i.e. 0.5m.

Question 2.11   The tide-producing force at P (TPF_P) = the Moon's gravitational attraction (F_{gP}) at P, minus the centrifugal force at P:

$$F_{gP} = GM_1M_2/(R - a\cos\psi)^2 \qquad \text{(equation 2.3)}$$

and the centrifugal force at P is the same as at all other points on Earth, and therefore equal to F_g at the Earth's centre, i.e. GM_1M_2/R^2 (from equation 2.1). So, by analogy with the reasoning leading to equation 2.2,

$$TPF_P = GM_1M_2/(R - a\cos\psi)^2 - GM_1M_2/R^2$$

Do not worry if you could not manage the subsequent algebra, but the expression simplifies to:

$$TPF_P = GM_1M_2 \frac{a\cos\psi(2R - a\cos\psi)}{R^2(R - a\cos\psi)^2}$$

giving the approximation:

$$TPF_P \approx GM_1M_22a\cos\psi/R^3$$

Question 2.12 None of the statements is true.
(a) False. The term 'syzygy' includes *both* conjunction and opposition. If the Moon is in opposition, it is also in syzygy. However, if the Moon is in syzygy it is not necessarily in opposition (i.e. it might be in conjunction).

(b) False. Spring tides would occur (see Figure 2.6(a)).

(c) False. Spring tides occur every 14.75 days throughout the year.

(d) False. The lowest sea-levels occur at low *spring* tides. When the Moon is in quadrature, neap tides occur, and these have a smaller range than spring tides.

Question 2.13(i) The equilibrium tides are constrained by the shallow depth of the oceans, and by the shape of the ocean basins.

(ii) There is a time-lag between the application of the tide-generating forces and the oceans' responses.

(iii) Tidal currents are subject to the Coriolis force.

Question 2.14(a) There would be a decrease in the tidal range. The difference between perihelion and aphelion in terms of Earth–Sun distance is only about 4% (Section 2.2), and the Sun has slightly less than half the tide-raising influence of the Moon. The decrease in tidal range as the Earth–Sun distance increased would therefore be small (say a centimetre or so).

(b) Changes in the Moon's declination, which cause equatorial and tropic tides (Section 2.1.1), have little effect at high latitudes. As Figure 2.11 shows, there is little diurnal inequality in the tidal range at Immingham (F-value = 0.1), although close inspection of the tidal curve does reveal very slight diurnal inequalities around days 6–8 and 21–23.

(c) No effect on tidal *range*, but by analogy to the answer to Question 2.10 a depression of the sea-surface of about 30cm would result, whatever the state of the tide at the time.

(d) The tidal range increases to about 6m.

Question 2.15 The South Pole. The Earth–Moon system is shown as rotating in a clockwise direction on Figure 2.1. Compare this with Figure 2.4, where both the rotation of the Moon about the Earth and the Earth's spin on its axis are shown as anticlockwise. The Earth's spin is towards the East, so also is the rotation of the Moon about the Earth, and therefore the rotation of the Earth about the Earth–Moon centre of mass is, likewise, towards the East. Figure 2.4 shows the North Pole. Therefore the pole shown in Figure 2.1 must be the South Pole.

CHAPTER 3

Question 3.1(a)   The vast drainage basin of the Amazon and its tributaries which flow across the wet, equatorial and tropical zones of South America is to the east of the Andes, and so river discharge into the Atlantic Ocean is high. This water carries much suspended sediment. There is little river run-off west of the Andes because the western coast of South America between the Equator and 30° S is arid. So, although extensive weathered debris is produced in the Andes, it cannot be transported easily by rivers westwards into the Pacific Ocean.

(b)   The south-west Pacific is also situated in humid equatorial and tropical climatic belts. This leads to high river discharge and weathering, and the transport of large quantities of sediment to the south-west Pacific Ocean.

(c)   The south-west Pacific Ocean is bordered by marginal basins behind island arcs. Much of the sediment discharged by rivers is trapped by deposition in these marginal basins. Any sediment discharged by rivers draining the island arcs on the oceanward side is deposited in the ocean trenches. So, little is available to reach the deep ocean basins. The Atlantic has no island arc and trench systems, except in the Caribbean Sea. Sediment discharged into the Atlantic is able to build up across the continental shelves where, eventually, it may cascade down the continental slope into the deep ocean basins.

Question 3.2   You should have selected (b), (c) and (e). Relict sediments are ubiquitous on the continental shelves, and volcanoes occur at all latitudes. However, it is also important to bear in mind that although recent glacial sediments are restricted to high latitudes, glacial material also occurs in relict sediments at mid-latitudes in the Northern Hemisphere. It was deposited when ice-sheets extended much further south during glacial periods in the Quaternary. Although carbonate sediments are most abundant at low latitudes, along tropical and arid coastlines, there are rare occurrences at higher latitudes where terrigenous sediment is scarce.

CHAPTER 4

Question 4.1   As the unit of force is the newton (N), and this force is being applied over an area of the bed, the units for shear stress are Nm^{-2}. Note that these are the units of pressure.

Question 4.2(a)   Very close to the bed, where frictional retardation is at its maximum, conditions will approach those of laminar flow. However, as velocity increases further away from the bed, the flow will become progressively more turbulent.

(b)   Close to the bed where the flow approaches laminar conditions, the viscosity is mainly the molecular viscosity of the water. However, as turbulence increases with increasing velocity, so the eddy viscosity will also increase, and so the viscosity effectively increases with increasing distance from the bed.

Question 4.3   With increasing current speed, the flow becomes more turbulent and the velocity profile becomes flatter towards the bed. This is because velocity increases more rapidly with height above the bed in a fast flow than a slow flow. Therefore, fully turbulent conditions will begin *closer* to the bed, and so the viscous sublayer becomes thinner.

Question 4.4(a)   The large grains in a coarse-grained sediment are more likely to disrupt the viscous sublayer than the grains in a fine-grained sediment; therefore, the flow roughness will increase, which increases the potential for sediment movement, even though the grains are on average larger.

(b)   Similarly, because the thickness of the viscous sublayer decreases as current speed increases, the grains are more likely to disrupt the sublayer as the flow speed increases. Once again, the roughness will increase, as will the potential for sediment movement.

Question 4.5   The base units of τ_0 (Nm^{-2}) are $(kgms^{-2}) \times m^{-2} = kgms^{-2}m^{-2}$ and those for ρ are kgm^{-3}. Therefore:

$$u_* = \sqrt{\frac{kgms^{-2}m^{-2}}{kgm^{-3}}}$$
$$= \sqrt{m^2s^{-2}}$$
$$= ms^{-1}$$

Question 4.6(a)   According to equation 4.6, $\tau_0 = \rho u_*^2$,
so $\tau_0 = (0.083ms^{-1})^2 \times 10^3 kgm^{-3}$
$= 0.0069 m^2 s^{-2} \times 1\,000 kgm^{-3}$
$= 6.9 kgm^{-1}s^{-2}$ or $6.9\ Nm^{-2}$.

(b)   The time-averaged current speed at 1m above the bed is $1.2ms^{-1}$ which is around two orders of magnitude greater than the fictitious shear velocity. This should give you some idea of real current speeds when you are considering shear velocity.

(c)   Over the interval 1m to $10^{-1}m$, $dz = \log_{10}1.0 - \log_{10}0.1$
$= 0 - (-1)$
$= 1,$
which is the same as over the interval 10^{-1} to 10^{-2}. As $d\bar{u}$ over the interval 1m to $10^{-1}m$ is almost the same as that over the interval $10^{-1}m$ to $10^{-2}m$ (1.2m − 0.71m = 0.49m), it follows that the shear velocity $d\bar{u}/dz$ must also be the same.

Question 4.7(a)(i)   If the bed roughness remained constant, then the roughness length, z_0, would stay the same; to accommodate an increase in speed, the slope would become flatter; in other words, the two slopes would diverge away from z_0 (see Figure A1).

(ii)   In this case, the roughness length would increase, so the intercept z_0 would occur higher up the depth axis. As the current speed is unchanged, the slope would, once again, become flatter; in other words, the two slopes would converge (see Figure A2).

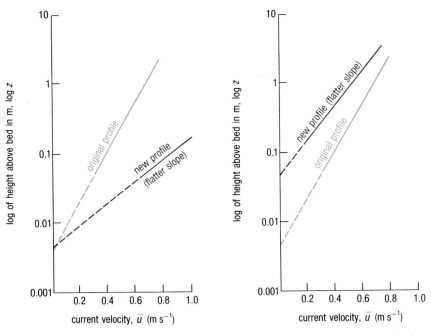

Figure A1 Logarithmic velocity profiles derived from Figure 4.6(b).

Figure A2 Logarithmic velocity profiles derived from Figure 4.6(b).

(b)   In both cases, a flattening of the slope implies a steeper velocity gradient. As $\tau_0 \propto d\bar{u}/dz$, this also means an increase in τ_0. Consequently, both an increase in current velocity and an increase in bed roughness will lead to an increase in shear stress and hence in erosion at the bed.

Question 4.8(a)   It is quite difficult to be precise measuring from a log scale on both axes, but you should have obtained a ratio somewhere between 1.0 and about 1.5. Experimentally, the ratio is found to be 1.2. Sediment coarser than 0.1mm includes everything from some very fine sand through to boulders, although it is highly unlikely that sediment the size of boulders would ever be lifted into suspension by natural flows in the sea.

(b)   You should have found this easier to measure, and have obtained a ratio of about 0.1.

(c)   These grains are lifted straight into suspension without entering a phase of bedload transport, although the coarser ones may be suspended intermittently, rather than being carried fully in suspension. Neither of the ratios applies. The ratio is less than 0.1, and becomes smaller with decreasing grain size.

Question 4.9(a)   The difference in velocity is about $0.03\,\mathrm{m\,s^{-1}}$, or $3\,\mathrm{cm\,s^{-1}}$, so there is only just over a 4% difference in velocity between the two tidal currents.

(b)   The 'SSW' shaded area in Figure 4.10(b) is about one-third as large again as the 'NNE' shaded area, and so there must be significant net south–south-westerly transport of sediment. This also means that the relatively small difference in current velocity has led to a 33% increase in sediment transport in one direction, because the shaded areas represent the total amount of sediment being transported.

(c)   The dashed lines represent the threshold velocity (in Figure 4.10(a), of the actual current, $\bar{u}$, and in Figure 4.10(b), of $\bar{u}^3$) required to move sediment grains of 0.3mm diameter. In Figure 4.10(b), the shaded areas represent those parts of the tidal cycle where this threshold is exceeded and so sediment grains coarser than 0.3mm are being moved, certainly in the bedload, and probably in suspension for part of the time. In the unshaded areas, where the curve falls below the threshold velocity, only grains smaller than 0.3mm are being moved, and the grain sizes being moved decrease as the velocity minima are approached.

Question 4.10   The coarser sediments (0.075 to 0.1mm) are concentrated in the bottom two or three metres of the water column (Figure 4.12(b)) and, despite the somewhat slower currents at these levels due to frictional retardation (Figure 4.12(a)), the sediment fluxes are greatest here (Figure 4.12(c)). In contrast, the finer-grained sediments are much more evenly spread through the water column, as you might expect, because the small particles are much more easily kept in suspension by turbulent flow. Concentrations rise only slightly in the bottom two metres and fall off in the top two metres. When concentrations are combined with the velocity distributions (Figure 4.12(a)), the minimum fluxes are at the top and bottom of the water column, a pattern very different from the one observed for the coarser suspended sediments.

Question 4.11   The settling velocity of grains 0.1mm in diameter is 10mms^{-1} and most of this sediment is concentrated within three metres (or 3×10^3mm) of the water column. Therefore it would take most of the sediment grains ($3 \times 10^3/10$) seconds to reach the bed, or about 5 minutes.

Question 4.12(a)   From the horizontal scale on Figure 4.9 you can read off the settling velocity of 0.01mm diameter grains as 10^{-1}mms^{-1}, or 10^{-4}ms^{-1}. The critical depositional shear velocity for grains this size is about 0.006ms^{-1}. From equation 4.6,

$$\tau_d = 1\,000 \times (0.006)^2 \text{ Nm}^{-2}$$
$$= 3.6 \times 10^{-2} \text{ Nm}^{-2}$$
and $$\tau_0 = 1\,000 \times (0.002)^2 \text{ Nm}^{-2}$$
$$= 4 \times 10^{-3} \text{ Nm}^{-2}.$$

You have already been told that C_b is 0.005kgm^{-3}. Substituting these values into equation 4.10,

$$R_d = (5 \times 10^{-3}\text{kgm}^{-3}) \times (10^{-4}\text{ms}^{-1}) \times \left(1 - \frac{4 \times 10^{-3} \text{ Nm}^{-2}}{3.6 \times 10^{-2} \text{ Nm}^{-2}}\right)$$
$$= (5 \times 10^{-7}) \times (1 - 0.11)\text{kgm}^{-2}\text{s}^{-1}$$
$$= 5 \times 10^{-7} \times 0.89\text{kgm}^{-2}\text{s}^{-1}$$
$$= 4.45 \times 10^{-7}\text{kgm}^{-2}\text{s}^{-1}$$

or 4.45×10^{-4} gm^{-2}s^{-1}, which does not seem very much.

(b)   The number of seconds in a non-leap year is $3\,600 \times 24 \times 365 = 3.1536 \times 10^7$, so the annual rate of deposition is $(4.45 \times 10^{-7}) \times (3.1536 \times 10^7)$, which is just over 14kgm^{-2} yr^{-1}—a very significant quantity.

(c)   If the current speed fell to zero, then the shear stress at the bed (τ_0) would be zero and the ratio between τ_0 and τ_d would also be zero. The value of the expression in brackets would be 1, so R_d would then depend only on the settling velocity and the concentration of sediment close to the bed, and would be equal to $5 \times 10^{-7} \text{kg m}^{-2}\text{s}^{-1}$. The consequent increase in rate of deposition when the current speed falls to zero is an important factor encouraging high rates of sediment deposition on estuarine and tidal mud-flats at the slack water of high tide.

Question 4.13(a)   Over the depth interval 10^{-1} to 10^{-2}m, the current velocity changes from about 0.38ms^{-1} to 0.1ms^{-1}, so dlog z is 1 and d$\bar{u}$ is 0.28ms^{-1}. From equation 4.8,

$$\bar{u}_* = \frac{\mathrm{d}\bar{u}}{5.75 \times \mathrm{dlog}\ z} = \frac{0.28}{5.75 \times 1}$$
$$= 0.048\text{ms}^{-1}.$$

From equation 4.6,

$$\tau_0 = \rho u_*^2$$
$$= (10^3) \times (0.048)^2\ \text{N m}^{-2}$$
$$= 2.3\ \text{N m}^{-2}.$$

The current velocity at one metre above the bed is about 0.65ms^{-1}, an order of magnitude larger than the calculated shear velocity.

(b)   As the current is a tidal current, it may be either accelerating or decelerating. This means that the values for u_* and τ_0 could be either underestimated or overestimated, respectively. Also, if the current has previously flowed over a bed with a markedly different roughness (e.g. gravel rather than sand), then the velocity profile may not yet have adjusted to take acount of this.

(c)   From Figure 4.9, the range of grain sizes that could be transported only as bedload is about 0.35mm to 4.5mm. This includes medium-grained sand to fine-grained gravel (Table 4.1).

(d)   The maximum grain size that could be transported fully as suspended load is about 0.07mm (70 µm), at the lower end of the very fine sand range.

(e)   The presence of bed forms would cause an added resistance to the flow (form drag). This reduces the capacity of the flow to move sediments. In other words, not all the shear stress calculated in (a) would be available to move sediment. Consequently, the answers to (c) and (d) would be overestimates.

Question 4.14   A current-produced bed form is asymmetrical with the steeper face pointing down-current. This means that the steeper face of any bed form should point in the direction indicated by the arrows on Figure 4.11.

Question 4.15(a)   Both the ripples and the megaripples are asymmetrical with the gentle slope pointing towards the top of the photograph, i.e. the top is downstream. This means that the current flowed from the bottom to the top of the photograph.

(b)   Generally, megaripples are formed at higher current speeds than small-scale ripples. Therefore, the megaripples could have formed when the current speed was high, and the superimposed small-scale ripples formed as the current speed waned.

(c)   Megaripples do not form in sediments with a grain diameter much finer than about 0.15 mm and small-scale ripples do not form in sediments much coarser than about 0.8 mm grain diameter (Figure 4.14). Therefore, since both types of bed form are present, the sediment grain size must be within the range 0.15 to 0.8 mm.

CHAPTER 5

Question 5.1   No, because of tidal action. On the flood tide, the zones will all move landwards and, conversely, on the ebb tide, they will all move seawards. The greater the tidal range, and the flatter the littoral zone, the further the zones migrate, as anyone who has attempted to walk down to the sea at beaches like that at Blackpool when the tide is out will know. In addition, wind-enhanced waves may drive the zones landwards.

Question 5.2(a)   The beach face coincides with the intertidal zone and can be considered synonymous with the foreshore.

(b)   The beach face is also the section of the beach profile exposed to the action of wave swash; however, it is not synonymous with the swash zone. The swash zone migrates up the beach with the tide and so at high tide much of the beach face may be in the breaker zone, whereas at low tide most of the beach face is exposed and only the seawards part is covered with swash.

Question 5.3(a)   Wave period remains constant as the frequency must be unchanged.

(b)   The wave speed decreases (Section 1.4.5).

(c)   If wave period remains constant but the wave slows down, then because the total wave energy remains the same, the wave height must increase (equation 1.14).

(d)   If wave height increases and wave depth decreases, then, according to equation 5.2, the maximum orbital velocity must also increase.

Question 5.4(a)   From Figure 5.4, the maximum grain size that could be moved is about 0.3 mm.

(b)   The answer is 'no'. From equations 5.2 and 5.3, if wave height decreases, u_m must decrease because u_m (and so u_t) $\propto H$; u_m will drop below the value of u_t required to move grains 0.3 mm in diameter.

(c)   Once again, the answer is 'no'. If water depth increases, u_m must decrease because $u_m \propto 1/\sqrt{d}$, and so will fall below the required value of u_t.

Question 5.5   All other things being equal, sediment grain sizes should become finer in a seawards direction, something you have probably seen for yourself on a beach near low tide.

Question 5.6(a)   If we insert the values given in the question into equation 5.4, then we have:

$$P_1 = 0.5 \ (1/8 \times 10 \times 9.8 \times 1) \times 0.5 \times 0.87$$
$$\approx 266 \ \mathrm{W\,m^{-2}}.$$

176

(b)   If the wave crest approaches parallel to the shoreline, then the angle between the wave crest and the shoreline must be zero. So, sin α and cos α must also be zero and the value of P_l will be zero as well, implying there is no *longshore* wave power. All the wave power must occur perpendicular to the shoreline.

Question 5.7(a)   For a current speed of about $0.35\,\text{ms}^{-1}$, and wave height 2m, the enhancement factor is approximately 9. In other words, wave action causes a nearly ten-fold increase in shear stress at the bed. When the current speed is halved, the enhancement factor is close to 10^2, in other words there is an order of magnitude increase in the enhancement factor.

(b)   The enhancement factor for a 2m wave is about 9 (see (a)), and for a 1m wave about 4 (remember that the vertical scale is logarithmic, not linear)—a difference of about 5. However, the difference in enhancement factor would begin to increase substantially as the current speed fell below $0.35\,\text{ms}^{-1}$.

Question 5.8(a)   Wave steepness is defined as the ratio of the wave height (H) to the wavelength (L), i.e. H/L.

(b)   In decreasing order of wave steepness, we have spilling, plunging, collapsing and surging breakers.

Question 5.9   The curve for grains of 3.44mm diameter (gravel-sized) indicates that the beach slope remains more or less constant at around 20° until a wave steepness of 0.02 is reached and then there is a sudden decrease in the angle of slope. This suggests that wave steepness is much less important than grain size in controlling beach slope until quite high values of wave steepness are reached. A similar, but not so marked, trend is seen in the curve for coarse sand-sized sediment (0.97mm). For the medium- and fine-grained sand-sized sediment, there is a continuous decrease in beach slope with increasing wave steepness. This is consistent with a reduction in the effects of percolation, where beaches are built of fine-grained sediment, and an increase in the influence of wave steepness.

Question 5.10   Waves increase in height in shallow water which leads to an increase in the orbital velocity (equation 5.3). This means that the enhancement factor should increase. Also, the decrease in water depth leads to a further increase in the orbital velocity, and enhancement factor.

Question 5.11   The speed of a longshore current is proportional to the maximum orbital velocities of the waves (equation 5.2). Because orbital velocities increase as wave height increases, large, high waves will generate stronger longshore currents than small waves.

Question 5.12   The actual outcome was reported in 1972 by the Jacksonville City Engineer, Oscar G. Rawls:

> 'The jetties trapped the southward-migrating sand and held it north of the north jetty. Sand accreted there. That accretion meant, however, that there was almost equal starvation of the beaches south of the jetties . . . and we have an eroding beach to the south of the jetties extending for several miles.'

CHAPTER 6

Question 6.1 The Wash forms a relatively narrow coastal embayment and so the coastline is sheltered from wave action. The north coast of the Netherlands is protected by the Friesian Islands which fringe the Wadden Sea. The coastline of the United Arab Emirates borders the sheltered Persian Gulf.

Question 6.2 Given the strong tidal currents, you would expect to see current-formed ripples, and even megaripples. There may also be wave-formed ripples.

Question 6.3(a) More seawater is likely to be drawn in beneath the mats to replace that being lost at the surface.

(b) As the seawater is drawn up through the algal mats to the surface, it will evaporate, precipitating salts within the algal mats to form evaporite deposits.

Question 6.4 The salt water that is lost must be replaced by an influx of seawater causing a slow, landwards flow of seawater beneath the freshwater (Figure 6.6(c)).

Question 6.5 The settling velocity of these particles is in the order of $10^{-3}\,\text{mm s}^{-1}$, or $10^{-6}\,\text{m s}^{-1}$. It would take them 5×10^{6} seconds, or around 1 389 hours (or nearly 58 days) to settle to the bed. Therefore, in theory, it should not be possible for muds to settle, even at slack water.

Question 6.6 The water speed will be reduced rapidly because the flow is no longer constrained to move through a small cross-sectional area and so the deposition from suspension of fine-grained sediments will begin.

Question 6.7 At the high spring tides, when tidal currents penetrate furthest up the estuary and are at their strongest, the turbidity maximum will also be at its furthest point up the estuary. It will also contain its maximum sediment concentration because of the abundant marine sediment brought up by the tidal currents, reinforced by the residual current. Conversely, at the high neap tide it will be at its furthest position down-estuary, and will have its minimum sediment concentration. (This answer assumes there are no major changes in the river discharge.)

Question 6.8 The result would have been an *increase* in the rate of sedimentation in the estuary because the landwards flow of saltwater at the bed would have led to an increase in the sediment load brought up the estuary.

Question 6.9(a) The surface isohalines are not straight, but are deflected so that they are pushed up-estuary further on the eastern side and pushed down-estuary on the western side, thus giving a lateral salinity gradient. This reflects the influence of the Coriolis force in the Northern Hemisphere, deflecting the tidal and river flows to the right.

(b) The isohaline pattern suggests that there is very strong tidal flow so that seawater enters the estuary on the eastern side, and river water flows down-estuary on the western side. It seems, therefore, to be a well-mixed estuary. In this type of estuary, there is no vertical water circulation but there may be a horizontal circulation between the two flows.

(c)   Seawater flowing up the eastern side of the estuary is likely to deposit marine sediments on the eastern bank whereas river-borne sediments are likely to be deposited on the western bank.

(d)   There is no evidence that the surface isohalines in the minor estuaries are deflected and so the water circulation pattern is not the same as that described in (b).

Question 6.10(a)   First of all, the maximum ebb tidal currents are much stronger than the maximum flood tidal currents. However, the maximum ebb tidal current eight hours into the tidal cycle is stronger than that recorded at approximately twenty-four hours.

(b)   Maximum shear stresses at the bed should occur when the tidal current speeds are at their greatest and will be higher at the maximum ebb current speeds than the flood current speeds. Because of frictional retardation with the bed, the current speeds at 6m depth close to the bed, should be retarded relative to the currents recorded at the surface. The fact that the maximum ebb tide speed at eight hours is much greater than that recorded at twenty-two hours suggests that the velocity gradient ($d\bar{u}/dz$) in the boundary layer developed at eight hours is much greater than that at twenty-two hours (i.e. $d\bar{u}/dz$ is greater), and so the shear velocity and the shear stress at the bed will also be greater.

(c)   There are peaks of suspended sediment concentration corresponding approximately to just after the maximum tidal current speeds, consistent with maximum erosion of the sediment bed at periods of maximum bed shear stress. As we should expect, the concentrations are greatest on the ebb tides, and more pronounced on the first ebb tide than the second.

CHAPTER 7

Question 7.1   The patterns are likely to be the same as those occurring in a salt wedge estuary. Denser, saline seawater penetrates the distributary mouth as a salt wedge over which the less dense freshwater flows (Figure 6.6(a)). The shear stress at the boundary between the seawards-moving freshwater and the underlying salt water generates internal waves along the boundary leading to the mixing of a small amount of salt water up into the freshwater layer. Seawater then moves landwards along the bottom of the distributary to replace the salt water lost by mixing.

Question 7.2(a)   Mixing of freshwater and salt water will occur at both the base and the sides of the plume.

(b)   The plume will expand over a wider front and the flow will decelerate.

(c)   The coarsest sediment will be deposited rapidly to form a bar at the mouth of the distributary (Figure 7.6(b) and (c)).

Question 7.3   You should recall (Section 6.2.2) that clay particles flocculate in salt water to form larger aggregates which settle much more rapidly than individual clay particles.

Question 7.4(a) If the basin is shallow, there is an obvious limit to how far the plume can expand vertically, and so there will be greater lateral expansion.

(b) Turbulent mixing will occur down as far as the bed because of the high speed and shallow depth.

(c) Shear stress at the bed just seawards of the distributary mouth will be significant because the residual flow of water will be seawards, and vigorous, down to the bed, as in a fully mixed estuary. This means that a large amount of coarse-grained bedload will be transported seawards from the mouth.

Question 7.5 The effects will be the same as in a well-mixed estuary.
(a) The density stratification will not become established and turbulent mixing will predominate. The residual flow will be downstream at all depths, but superimposed on this will be a landwards flow associated with the flood tide and a seawards flow associated with the ebb tide.

(b) Sediment will not only be moved downstream as a result of the residual flow, but will also be moved both landwards and seawards by the flood and ebb tides.

(c) Because of daily and monthly variations in the positions of mean high water and mean low water, the interface between fluvial and marine processes will periodically shift up and down the distributary channels.

Question 7.6(a) The wave speed will decrease.

(b) The wavelength will decrease.

(c) The wave height will increase.
If you had difficulty in obtaining these answers, then you should read Section 1.5.1 again.

Question 7.7(a) The conditions under which density stratification is most likely are those outlined in (iv). All other conditions will probably lead to turbulent mixing sufficient to completely or partially prevent the density stratification.

(b) The conditions most likely to lead to a fully mixed distributary mouth are those outlined in (i) because turbulent mixing occurs down to the sea-bed.

Question 7.8 According to Figure 7.12, wave action and tidal currents dominate the Copper delta equally. The delta shows the ragged outline and longitudinal sediment ridges characteristic of a tide-dominated delta. However, at the seawards edge the sediment has been reworked into a series of linear islands parallel to the shoreline, so the influence of waves is also apparent. The Fly delta shows the ragged outline, the longitudinal sediment ridges, and the funnel-shaped distributary mouth characteristic of a completely tide-dominated delta. The Mahakam delta shows some evidence for distributaries with long, straight finger-like deposits associated with river-dominated deltas. However, some of the distributaries also contain longitudinal sediment ridges, and so it seems reasonable to infer that both fluvial and tidal processes influence the Mahakam delta.

Question 7.9(a) The Nile delta shoreline is relatively straight with few distributaries, and the photograph shows how the sediment has been reworked into beaches. The dominant influence appears to be wave action. However, two distributaries discharge into the sea from quite elongate protuberances which suggests that fluvial processes also play a part in shaping the delta. The Nile should, therefore, plot in the wave-dominated sector of Figure 7.12, but part-way towards the river-dominated sector.

(b) The Po delta shows some features in common with the Nile delta. The movement of sediment back landwards to form beaches is visible, suggesting wave action is important. However, the delta protrudes into the sea far more than the Nile does, and the delta plain is crossed by a larger number of distributaries, so fluvial processes are even more significant. The Po should, therefore, plot in the river-dominated sector, but part-way towards the wave-dominated sector.

CHAPTER 8

Question 8.1(a) Physical weathering and the mechanical shattering of rocks are likely to produce gravels and coarse sands.

(b) In the absence, or virtual absence, of terrigenous sediments, carbonates can accumulate. Carbonate sediments may be composed of muds, fragments of shell or coral, and occasionally grains which have been precipitated inorganically from seawater.

Question 8.2 The coarse-grained gravels are likely to disrupt the viscous sublayer, increasing flow roughness and allowing turbulence to occur right down to the bed. This increases the potential for sediment erosion. A viscous sublayer is developed over the silt, however, and so any fine particles being transported near the bed will settle into the sublayer once the bed shear stress falls below the critical depositional shear stress.

Question 8.3(a) Decreasing water depth will cause an increase in the wave height, so the amplitude will increase.

(b) If the amplitude increases, then the tidal range will also increase.

(c) The tidal current speed will also increase.

Question 8.4 The highest tidal current ranges are found off the coast of Argentina and at the mouth of the Amazon river; and it is there that the shelf widths are greatest. However, you should also note that these stretches of coastline have marked coastal embayments as well, which may increase the possibility of resonance.

Question 8.5 The rate of bedload transport, q_b, is proportional to the cube of the shear velocity, u_*^3 (equation 4.9), and so is proportional to $\bar{u}^3$, the cube of the average current speed. A small difference in current speed will, therefore, lead to a large difference in sediment transport rates.

Question 8.6(a) Bedload partings coincide with those areas where the maximum sea-bed shear stress attains the highest values in opposing directions, which is what you would expect. On the whole, there is good agreement between the directions of sand transport shown in Figure 8.4 and the directions of the bed shear stress vectors. There are some areas of disagreement, however; e.g. the simulation suggests weak sediment transport to both the east and the west of a parting in the western English Channel. (In fact, the asymmetry of the bed forms in this part of the Channel shows that all the net sediment transport is westerly.)

(b)   From Figure 8.5, the maximum bed shear stress is about 7 Nm^{-2} (in the Bristol Channel). Using equation 4.5, the shear velocity, u_*,

$$= \sqrt{\tau_0/\rho}$$
$$= \sqrt{7/(1 \times 10^3)}\,\text{ms}^{-1}$$
$$\approx 0.084\,\text{ms}^{-1}.$$

According to Figure 4.9, the maximum grain size that would be moved is gravel around 8mm in diameter. However, in practice a considerable proportion of the bed shear stress will be due to form drag (Section 4.1.6), and so be unavailable for sediment movement.

Question 8.7(a)   The wave height should be about 2.6m.

(b)   Figure 5.4 applies to sediment movement on a flat bed. It is highly unlikely that the sea-bed will be flat; it is likely to be disturbed by bed forms. Consequently, only a small proportion of the bed shear stress calculated from the wave orbital velocity will be available for sediment movement.

Question 8.8   As the sea-bed is not flat, but is disturbed by the sand-waves, the local bed shear stress available for sediment transport is the total shear stress *minus* the form drag, and may be within the range for ripples or megaripples to develop.

Question 8.9   Dredging is likely to disturb the dynamic equilibrium of the littoral zone, leading to extensive erosion of adjacent coastal areas to replace the removed sediment.

Question 8.10(a)   There is a direct relationship between the two in general terms. Exposed rock surfaces or gravel are found where the tidal currents are greatest, usually in excess of 1ms^{-1}. With decreasing current speeds, coarse sand occurs, followed by fine sands. Muds seem to be restricted to regions where the current speeds are less than about 0.4ms^{-1}.

(b)   Sand-ribbons occur where the current speeds are greater than about 0.8ms^{-1} and sand-waves are found at current speeds between about 0.6 and 0.8ms^{-1} (see also Figure 8.6).

(c)   Broadly speaking, sediment grain sizes decrease along the directions of the sediment transport paths and there is also a progression from sand-ribbons to sand-waves. This is what we should expect from the sediment transport paths in Figure 8.4.

(d)   Furrows should occur where the current speeds exceed 1.5ms^{-1}, in other words at the zones of bedload parting in the English Channel (see Figure 8.4).

182

Question 8.11   Scour hollows are elongate features, approximately parallel to the direction of current flow. Therefore, the tidal currents responsible for the scour hollows shown must be directed approximately SE–NW through the Straits.

Question 8.12   Placer deposits can be regarded as residual sediments in the sense that fine-grained and less dense particles have been winnowed out by wave and current action leaving the heavy minerals behind as residual deposits. However, some are also relict river or beach sediments, formed during periods of lower sea-level.

ACKNOWLEDGEMENTS

The Course Team wishes to thank the following: Dr. Martin Angel and Dr. Chris Vincent, the external assessors, both of whom also provided helpful advice on content and level; Maurice Crickmore and Mary Llewellyn for advice and comment on Chapters 3–6; and Mike Hosken for advice and comment on the whole Volume.

The structure and content of this Volume and of the Series as a whole owes much to the experience of producing and presenting the first Open University Course in Oceanography (S334), from 1976 to 1987. We are grateful to the Course Team members who prepared and maintained that Course, to the students and tutors who provided valuable feedback and advice, and to Unesco for supporting its use overseas.

Grateful acknowledgement is also made to the following for material used in this Volume:

Table 2.1 and Figure 1.2 H. J. McLellan (1965) *Elements of Physical Oceanography*, Pergamon; *Figures 1.4 and 2.5* J. G. Harvey (1976) *Atmosphere and Ocean*, Artemis Press; *Figures 1.11 and 1.14* B. Kinsman (1965) *Wind Waves*, Prentice-Hall; *Figure 1.17* C. E. Vincent; *Figures 1.19, 1.20, 7.1, 7.3, 7.5 and 7.9* NASA; *Figure 2.9* The Royal Society; *Figure 2.10(a–c)* H. U. Sverdrup *et al.* (1942) *The Oceans*, Prentice-Hall; *Figures 2.11 and 2.12(b)* A. Defant (1961) *Physical Oceanography*, Pergamon; *Figure 2.14* F. T. Banner *et al.* (eds) (1980) 'The North-West European Shelf Seas' in *Elsevier Oceanography Series 24B*, Elsevier; *Figure 2.15(a)* B. J. Skinner and K. K. Turekian (1973) *Man and the Ocean*, Prentice-Hall; *Figure 2.15(b)* French Embassy, London; *Figure 3.1* J. D. Milliman and R. H. Meade (1983) in *Journal of Geology*, **91**, University of Chicago Press; *Figures 4.8, 4.15 and 6.13* K. Dyer (1986) *Coastal and Estuarine Sediment Dynamics*, Wiley; *Figure 4.9* I. N. McCave and the Geological Society; *Figure 4.11* I. N. McCave (1971) in *Marine Geology*, Elsevier; *Figure 4.14* L. A. Boguchwal; *Figure 4.16(a) and (b)* J. R. L. Allen (1985) *Principles of Physical Sedimentology*, Allen & Unwin; *Figures 5.1, 5.5, 5.6(a) and 5.13* P. D. Komar (1976) *Beach Processes and Sedimentation*, Prentice-Hall; *Figure 5.2(a)* Dee Edwards; *Figure 5.3* Elsevier Science Publishers; *Figures 5.4 and 7.4* Society for Economic Palaeontologists; *Figures 5.6(b) and 6.12* J. V. McCormick and J. M. Thiruvathukal (1985) *Elements of Oceanography*, CBS College Publishing; *Figure 5.7* J. Pethick (1984) *An Introduction to Coastal Geomorphology*, Edward Arnold; *Figure 5.11* American Society of Civil Engineers; *Figures 6.1 and 6.5* National Remote-Sensing Centre, Farnborough; *Figure 6.2(b)* Angela Colling; *Figure 6.3(b)* Jonathan Silvertown; *Figure 6.4(a)* J. B. Wright; *Figure 6.10* Devon Association for the Advancement of Science, Literature and Art; *Figure 6.11* E. Olavsson and I. Cato (1980) *Chemistry and Biogeochemistry of Estuaries*, Wiley; *Figures 7.6–7.8 and 7.10* L. D. Wright (1977) in *Geol. Soc. Amer. Bull.*, **88**; *Figure 7.12* W. E. Galloway (1983) *Terrigenous Clastic Depositional Systems*, Springer-Verlag; *Figure 8.1(a)* J. D. Milliman (1974) *Recent Sedimentary Carbonates, Part I*, Springer-Verlag; *Figure 8.1(b)* British Petroleum Development Ltd.; *Figure 8.2* D. G. Smith (ed.) (1982) *The Cambridge Encyclopaedia of Earth Sciences*, Cambridge University Press; *Figures 8.4, 8.6–8.9* A. H. Stride (ed.) (1982) *Offshore Tidal Sands*, Chapman & Hall; *Figure 8.11* Insitute of Oceanographic Sciences.

INDEX

Note: page numbers in italic refer to illustrations

accelerometer, wave measurement by 37
aggregates 156-158
aggregation, biological, in estuaries 122
Agulhas Current
  giant waves in 34
  shelf processes related to 146
air resistance 24
Amazon
  bore 63
  delta plain 130
  turbulent mixing in outflow 135
amphidromic point 55
amphidromic systems 55
amplitude, waves 7
angular frequency 17
antidunes 91, *91, 92*
antinodes, waves 35
aphelion 50
apogee 49
attenuation of wave energy 24-25

backshore 95
backwash, sedimentary structure formed by 96,
  108
Bay of Fundy, tidal range in 60,148
beach face 97
beaches 95-111
  materials making up sedimentary structures in
    108-109
  profiles 106-107
  relation to wave type 106
Beaufort Scale 13
bed forms 155-156
  continental shelf 153-156
  in relation to current flow 90-92
  in relation to grain size 90-92
bedload 67
  deposition 87
    transport rate 84-86
bedload partings 151
berm 96
biological aggregation 122
Blake Plateau 146
body waves 9
bores, tidal 62, 63
boundary layer 73, 74
breaker zone 96
breakers 29
Bristol Channel
  dredging 158
  tidal currents, shelf processes 148

calcium carbonate, sea-bed resources, dredging
  158
calcium phosphate, offshore dredging 159
Canary Current, shelf processes 146
capillary waves 9
carbonates
  beaches 108
  sediments, shelf 144, 158
  tidal flats, low-latitude 115
cassiterite, offshore, dredging 158

chemical weathering, shelf sediments 144
Chernobyl, contamination from, in sediments
  68
Chesapeake Bay 127, *127, 128*
clay minerals 67
coal, beaches 108
coastal erosion 68
cohesion, sediments 81
collapsing breakers 29
computers, tides, prediction 58
conjunction, Sun–Moon, spring tides 50
continental shelf, bed forms 153-156
convergence, deltas 133
coral, beaches 108
Coriolis force
  estuaries 120
  shelf processes 146
  tides, dynamic theory 52
counter-currents 32
co-range lines 55
co-tidal lines 55
critical depositional shear stress 87
critical shear stress 81
critical shear velocity 81
current flow, bed forms 90, 92
current ripples 92
current shear 60
  at the sea-bed 76-78
current speed, tidal 149
currents
  bed forms 92
  interaction with waves 31-32
  ocean, shelf processes 146
  sediment movement by 72-94, 105-106
  tides, shelf processes 148-152
  wave-generated, sediment transport 101-106
  wind-driven 146

Danube estuary 119
declination of Sun and Moon 48-49
deep water wave speed 17-19
delta bar *135*
delta front 130
delta plain 130
deltas 129-143
  river-dominated 132-137
  structure 130-131
  tide-dominated 137-139
  wave-dominated 139-140
density stratification, deltas 133
deposition of sediment 87-89
  bedload 87
  critical shear stress 87
  in distributary mouths 132-141
  on tidal flats 112-116
  rate 88-89
  suspended load 87-88
diamonds, offshore dredging 158
direct tides 55
dispersion, waves 19-21
displacement, waves 14
dissipation, wave energy 24

distributaries, deltas 130
distributary mouths
  mixing 132-141
  sediment deposition 132-141
distribution, sediments, tidal flats 112-115
diurnal inequalities, tide 58
diurnal tides 58
divergence 133
double high water, Southampton 59
dredging 157-159
  aggregates 157
  Great Bahama Bank 158
  Gulf of Mexico 158
  Iceland 158
  United Kingdom 158
drift, waves 15
dynamic balance of estuaries 125-126
dynamic theory, tides 52-58

Earth–Moon system, tide-producing
  forces 43-49
Earth–Sun system, tide-producing
  forces 50-52
East China Sea 144
ebb tides 43
eccentric rotation, Earth 43
eddies 74
eddy viscosity 74
energy
  tides 63
  waves 8
English Channel
  sediment, shelf processes 151
  shelf sea 144
epeiric seas 144
epicontinental platforms 144
equatorial tides 49
equilibrium tide 47
erosion of sediments 81-83
  cohesive 83
  non-cohesive 82-83
estuaries 112-128, 116-126
  dynamic balance 125-126
  negative circulation pattern,
    sedimentation 125
  partially mixed 119
  salt wedge 118
  sedimentation 122-123
  tides 62
  types 116-122
  well-mixed 120
evaporative pumping 116

fetch 12
Firth of Forth 120
flocculation 122
flocs 83
Florida Current 146
flow in a fluid 72
  laminar 74
  turbulent 74-76
foreshore 95

form drag 80
frequency of waves 7, 14
friction 60
  boundary layer 73
Friesian Islands 112
fronts 60, 133
fully developed sea 12
furrows 154

Ganges–Brahmaputra
  delta 131, 138
  delta plain 130
giant waves 32-34
Gironde, estuary 120
Goodwin Sands 156
grain size, effect on beach profiles 106
gravel, offshore dredging 157
gravity waves 9
group speed, wave dispersion 19-21
Gulf Stream 146
Gulf of Carpentaria 144

halocline 118
harbours, estuaries, sedimentation of 125
harmonic method, tides, prediction of 57-58
Hayasui Strait, scour hollows 154
high tidal flats 114
highest astronomical tide 52
Hudson Bay 144
Humber estuary 117, 120
  dredging 158
  sediment 139
hydraulic currents 60
hyperbolic tangent 17,18

ice-borne sediments 68
ilmenite, offshore dredging 158
indirect tides 54
interference, waves 19
internal waves 9
intertidal zone 95
Irish Sea
  scour hollows 154
  sediment 151

James River 120
John o' Groats, shelf sediments 144

Kelvin waves 57

ladder-back ripples 108
lag, tides 48
laminar flow 74
levées 133, *136*
littoral zone 95-111
  divisions 95
  sediment 96, 98-106
Liverpool Bay
  dredging 158
  sediment 151
logarithmic velocity profile 78, 80
longshore bar 96
longshore currents 101
longshore sediment transport 101, 104-105
low tidal flats 114
low-latitude tidal flats 115-116
lunar day 48
lunar tides, interaction with solar tides 50

Lynmouth 108

mangrove swamps 115
mangrove trees 115
maximum wave height 13
megaripples 90, *91*, 92, 155
Mersey estuary 120
  flocculation in 123
  silting up 126
mid-flats 114
Mississippi 119, 133, 137
  delta plain 130
mixing
  in distributary mouths 132-141
  in shallow seas 60
  between fresh water and seawater 116-121,
    135
molecular viscosity 74
monazite, offshore dredging 158
Moon
  declination 48-49
  elliptical orbit 49
mud-flats 112, 113, *113*, 114,
  114, 115

neap tides 50
negative estuarine circulation 121
  sedimentation 125
negative storm surge 61
Netherlands, tidal flats 112
Niger delta, mangrove swamps 115
Nile
  delta 129
  estuary 119
nodes, waves 35
non-cohesive sediments 81, 82
non-linear wave–wave interaction 24
North Sea
  bedload sediments, transport 85
  shelf sea 144
  storm surges 61
notation 5-6
null point 119

oblique waves 101-102
ocean currents 146
offshore movement, sediment 99-101
offshore zone 95
onshore movement, sediment 99-101
opposition, Sun–Moon, spring tides 50
orbital velocities, waves, sediment movement
  98-99
Ord delta, tides 137
Orford Ness, tidal flats 112

partial tides 57
partially mixed estuaries 119
  sedimentation 124
particle motion, waves 15
perigee 49
perihelion 50
period
  tides 57
  waves 7, 14
phase
  tides 57
  wave speed 19
phosphorites

offshore dredging 159
  sea-bed resources 156
photography, waves, measurement 39
physical weathering, shelf sediments 144
placer deposits 158
  sea-bed resources 156
planetary waves
  see Rossby waves
plunging breakers 29
Po estuary 119
pororoca 63
ports, estuaries, sedimentation 125
positive storm surge 61
prediction, tides 57-58
pressure gauges 37
prodelta 131, *131*
progressive waves 8

quadrature, neap tides 50
quartz 67
Quaternary glaciations 69

radar altimetry 37, 38
Rance estuary, tidal power 63
reefs, aggregates, dredging of 157
refraction, waves 25
relict sediments 70
remote sensing, waves 39
residual currents, estuaries 119
resonance 148
resonant period 35
rhomboid pattern 108
Rhône estuary 119
Rio de la Plata estuary 120
rip currents 102
ripples
  bed forms 155
  current-formed 108
  wave-generated 108
rivers
  sediments 68
  tides 62
river-dominated deltas 132-137
Rossby waves 10
rough turbulent flow 76
roughness, sea-surface 23
roughness length 78-80
runnels 96
rutile, offshore dredging 158

sabkha 116
Salicornia, tidal flats 115
shallow-water sediments, distribution
  of 69-70
salt wedge estuaries 118
  sedimentation 123-124
San Francisco Harbour, scour hollows 154
sand, offshore dredging 157
sandbanks 156
sand-ribbons 154
sand-waves 155-156
satellites 37 *see also* radar
  altimetry
Savannah River estuary, silting up 126
scatterometry 38
scour hollows 154
scouring 146
sea-bed

current shear 76-78
  forms, sediments 90-92
sea-bed resources 156-159
sedimentation in estuaries 122-123
sediment bar *136*
sediments
  cohesive 81, 83
  deposition 87-89
    in distributary mouths 132-141
    rates 88-89
    on tidal flats 112-116
  erosion 81-83
  littoral zone 96, 98-106
  longshore transport 101-104
  movement
    in bed forms 90-94
    theoretical background to 72-90
  non-cohesive 81, 82
  offshore movement 99-101
  onshore movement 99-101
  shallow water 67-72
    distribution 69-70
    supply 67, 69-70
  shelf 144-146
  structures, beach materials 107-109
  transport 84-87
    longshore 104-105
    tidal flats 112-116
    waves/currents 105-106
    see also sediment movement
  transport paths 150
seiches 35-37
seismic P-waves 9
seismic S-waves 9
semi-diurnal tides 50
settling lag 88
settling velocity 82
Severn
  bore 63
  estuary 120
shallow seas
  tidal currents 59-61
  tides 59-61
shallow water environments
    sediments 67-71
  changes with time 70
shallow water wave speed 17-19
shear stress 73 77
  at sea-bed 98-99
  critical depositional 87
  currents 149
  turbulent 74
shear velocity 78
shearing force 73
shelf
  processes 146-153
  sea-bed resources 156-159
  sediments 144-146
shelf seas 144-160
shell material, dredging of 158
shells, beaches 108
sheltering effect 10
Shoalhaven River 140
shore, effect of, on waves 27-31
shoreface 95
significant wave height 13
silting up
  harbours 125

ports 125
simple harmonic motion 9
Sitka Sound 144
smooth turbulent flow 75
solar tides 50
Solway Firth 151
sound waves 9
Spartina, tidal flats 115, 115
speed
  definition 5
  waves 17, 19
spilling breakers 29
spring tides 50
St George's Channel 151
standing waves 8
standing-wave tide 148
steepness, waves 7
storm surges 61-62
  negative 61
  positive 61
  shelf processes 152
Straits of Dover, bedload convergence 157
stratification between freshwater and sea
    water133
surf zone 96
surface tension 9
surface wave theory 17
  assumptions 19
surface waves 9
surging breakers 31
Surinam, tidal flats 112
suspended load
  deposition 87-88
  transport 86
swash bars 96
swash zone 96
swell 22-24
synthetic aperture radar (SAR) 38
syzygy 50
Tambora volcano 68
Thames estuary 120
  dredging 158
threshold velocity 99
tidal bores 62
tidal current speeds 149
tidal currents
  shallow seas 59-61
  shelf processes 148-152
tidal flats 112-128
  high 114
  low 114
  low latitude 115-116
  mid 114
tidal power 63-64
tidal ranges 149
tide-dominated deltas 137-139
tides 43-66
  deltas 137-139
  dynamic theory 52-58
  estuaries 62
  lunar-induced, variations 48-49
  prediction, harmonic method 57-58
  rivers 62
  shallow seas 59-61
  types 58-64
tide-producing forces 43-49
  Earth–Sun system 50-52
tractive force 46

transport of sediments 84-87
  by waves/currents 105-106
  in bed load 67, 84-87
  in suspended load 86-88
tropic tides 49
tsunamis 34-35
turbidity maximum, partially mixed
    estuaries 124
turbulent flow 74
  rough 76
  smooth 75
turbulent mixing 135

van der Waals forces 122
velocity, definition 5
velocity gradient 77
velocity profiles 80, 81
  in boundary layer 73
viscous attenuation, waves 24
viscous sublayer 75-76
volcanic eruptions 68
von Karman constant 78

Wash
  tidal currents 148
  tidal flats 113
Washington–Oregon shelf 148, 152
water particles, waves, motion 15
wave action 152-153
wave dispersion 19-21
wave displacement 14
wave drift 15
wave energy 2-31
  attenuation 24-25
  effect of shore line 27-31
  harnessing 22
  propagation 2-22
wave field, fully developed sea 12
wave height 7, 14
wave motion 9
  characteristics 8
wave number 17
wave power 22
wave refraction 25-27
wave set-up 102
wave size, fully developed sea 12
wave speed 17
  deep water 17-19
  shallow water 17-19
wave steepness 14
wavelength 7
waves 7-42
  definition 8
  interaction with currents 31-32,
    101-106
  measurement 37-39
  oblique, longshore currents 101-102
  offshore movement, sediment 99-101
  onshore movement, sediment 99-101
  orbital velocities 98-99
  satellite observation 37
  sediment movement 72-94
  types 8-10, 106
  unusual character 31-37
  wind-generated, ocean 10-11
wave-dominated deltas 139-140
wave-forms 14-19
wave–wave interaction 24

weathering
  chemical 144
  physical 144
well-mixed estuaries 120
  sedimentation in 125

white-capping 11, 24
wind speed 11
wind stress 146
yield strength, sediments 83